THE
CIRCADIAN
CODE

THE
CIRCADIAN
CODE

Lose Weight,
Supercharge Your Energy,
and Tranform Your Health
from Morning to Midnight

SATCHIN PANDA, PHD

All rights reserved.
Published in the United States by Rodale Books, an imprint of the Crown Publishing
Group, a division of Penguin Random House LLC, New York.
crownpublishing.com
rodalebooks.com

RODALE and the Plant colophon are registered trademarks of
Penguin Random House LLC.

This book is intended as a reference volume only, not as a medical
manual. The information given here is designed to help you make
informed decisions about your health. It is not intended as a substitute
for any treatment that may have been prescribed by your doctor. If
you suspect that you have a medical problem, we urge you to seek
competent medical help.

Library of Congress Cataloging-in-Publication Data
is available upon request.

ISBN 978-1-63565-243-7
Ebook ISBN 978-1-63565-244-4

Printed in the United States of America

Book design by Nancy Singer
Illustrations by Satchin Panda, PhD
Jacket design by Amy C. King

10 9 8 7 6 5 4 3 2

To my loving grandparents,
Banchhanidhi and Urbashi Panda,
Kalpataru and Leelabati Otta

Contents

Preface

To have rhythm, to be in sync, is to be healthy.
But not just any rhythm will do.

Germ theory—and its related breakthroughs of sanitation, vaccination, and antibiotics—was the groundbreaking health development of the past century. It prevents infectious disease and led to the most dramatic rise in longevity in any century in human history. Yet living longer does not always mean living healthier. In fact, we are now witnessing a rapid increase in chronic diseases of both the mind and the body beginning in early childhood and stretching through old age. Luckily, we are beginning to understand the cause: Our modern lifestyle is disrupting a deeply ingrained, primordial, and universal code to being healthy.

The observations that I have made over the past twenty years, along with my colleagues and other researchers in the tiny field of circadian biology, are radically changing the way we understand how both the body and the mind optimally function. The science of circadian rhythms is actually a multidisciplinary field that includes biologists, exercise physiologists, mathematicians, psychologists, sleep researchers, nutritionists, endocrinologists, ophthalmologists, geneticists, oncologists, and more. Working together, we have found that simply adjusting the timing of how we live—and making easy lifestyle changes—is the secret to restoring our rhythm, and it will surely be the next revolution in health care. I invite you to learn what I have uncovered through my own research and through working with the best minds in each of these fields. I call this the *circadian code*, and by adopting these lessons you will make small changes to the way you sleep, eat, work, learn, exercise, and light up your home that will make a profound difference in every aspect of your health. In fact, the benefits you'll reap can be far more effective and long-lasting than any medication or special diet.

You may have heard about circadian rhythms before; a 2017 Nobel Prize recognized this field of research for its impact on human health. But if you haven't heard of circadian rhythms, don't worry; the concept is very simple. The term *circadian* comes from the Latin *circa*, meaning "around" (or "approximately"), and *diēm*, meaning "day." Circadian rhythms are real biological processes that every plant, animal, and human exhibits over the course of a day. These rhythms are actually interconnected among species and are governed by internal circadian or biological clocks, which are very different from the "ticking biological clock" you might think about if you're worried about having kids by a certain age. As you'll learn, almost each and every one of our cells contains one of these clocks, and each is programmed to turn on or off thousands of genes at different times of the day or night.

These genes influence every aspect of our health. For instance, when we are healthy, we can have a good night's sleep. In the morning, we wake up feeling fresh and energetic and ready to get to work. Our gut function is perfectly normal. We have a healthy hunger and a clear mind. In the afternoon, we have the energy to exercise. At night, we are tired enough to go back to sleep without much effort. Yet when these daily rhythms are disturbed for as little as a day or two, our clocks cannot send out the right messages to these genes, and our body and mind will not function as well as we need. If this disruption continues for a few days, weeks, or months, we may succumb to all types of infections and diseases, ranging from insomnia to attention deficit hyperactivity disorder (ADHD), depression, anxiety, migraine, diabetes, obesity, cardiovascular disease, dementia, and even cancer.

Luckily, it's easy to get back in sync. We can optimize our clocks in just a few weeks. By restoring our circadian rhythms, we can even reverse some of the diseases or accelerate cures, returning us to better health.

My Journey: Discovering the Secrets of the Biology of Time

I was lucky to be born (in 1971) and raised in India during a unique time in history. I experienced firsthand how a rapidly evolving modern society disrupts the interconnectedness of life, including our own biological rhythms. Throughout my early childhood, I lived in a small town near my maternal grandparents. My grandfather worked as a goods clerk at the local train station, where he often worked the night shift. My grandparents lived in a house with a large jasmine tree near their front gate. To me, that tree was magical: It bloomed profusely at night and would shed its flowers right before dawn, as if laying a beautiful carpet to welcome my grandfather home each morning.

During summer and winter breaks, we visited my father's family, who lived on a farm in a rural part of the country. The contrast between my maternal grandfather's shift work at the train station and my paternal grandfather's life on the farm, in sync with nature, seemed like it was at least a century apart, although it was only a 2-hour road trip to get from one place to the other. For most of my childhood their village did not have electricity, so as you can imagine, life on the farm was very different from life at my home. My relatives raised almost everything they ate. Although I don't remember my grandfather ever wearing a watch, their daily routine happened with clockwork precision that was in sync with the sun and the stars. At the crack of dawn roosters sounded the alarm clock that woke everyone up. The whole day was spent tending to plants and animals and preparing meals. We picked fruits and vegetables or helped my uncle catch fish from the farm pond. Breakfast and lunch were the main meals, and they were feasts prepared with freshly picked vegetables and fish. Dinner was always before sunset and was mostly leftovers from lunch, as it was impossible to store any cooked food overnight. The evenings were also very different. The only light we had available came from kerosene lanterns. In those days kerosene was expensive and rationed by the government. My grandparents had a relatively large house with six bedrooms. We were only allowed to use the lanterns for a couple of hours in the evening, except for two lanterns that were placed at the ends of the verandah that flickered all night. After dinner all of the kids would huddle around a lantern, and my mother—who was a schoolteacher—would quiz us. Sometimes our aunts would join to tell us stories or my uncle would take us to the backyard to teach us the phases of the moon.

I remember asking for certain fruits or vegetables I liked to eat at home and being met with a strange look from my cousins. To them, I was a dumb town kid who didn't know which fruits and vegetables grow in what season. But what they didn't know was that my father, who had a college degree in agriculture, had introduced many high-yielding trees, vegetables, and rice varieties to my grandfather's farm. Some of these new rice strains could even grow in both the summer and winter, essentially doubling the return from the same piece of land. In this instance, disrupting the natural order of things didn't seem like such a bad idea.

When I was in junior high school, I lost my father to a road accident. A truck driver who was most likely sleep-deprived lost control of his vehicle. Years later I learned a sleep-deprived brain is more dangerous than a brain under the influence of alcohol. Yet even today, driving after a sleepless night is not illegal.

After high school, I went to an agriculture school like my father had, which at the time was the fastest path to a secure job in government or banking. Whenever I visited my grandparents' rural village, my grandfather would tease me and ask if I could crack the code of nature so that he could grow any fruit or vegetable in any season. That is how I developed an interest in understanding how all living things connect to daily and seasonal time.

I would also visit my maternal grandfather, who by then had retired from his job. It was only a few years after retirement that he started showing signs of dementia. My grandmother took care of him like he was a baby. I visited him almost every weekend in my senior year: I was one of only three or four people he recognized. He lost sense of day and night; he would feel hungry, sleepy, or could stay wide awake at random times. I began to notice how important the simple code of time is in our daily life. A few days after I completed college, he passed away at the age of 72.

I had done well in college with a major in plant breeding and genetics. My natural next step would have been to go for a master's degree in these same subjects, but I was lucky to secure a scholarship for a master's degree in molecular biology, which in India is called *biotechnology*. Molecular biology was at the time a new branch of science, and it introduced me to the genetic code.

Afterward, I landed a nice research job in the city of Chennai with Bush Boake Allen (now International Flavors and Fragrances), which makes flavoring agents and fragrances for almost every major food company in the world. My first assignment was to figure out the chemistry of how vanilla beans get their flavor. I visited the vanilla farms in the Nilgiri Hills in southern India, where my host would wake me up around 2:00 a.m. to drive to the fields to show me how the workers pollinated each vanilla flower by hand as soon as it would open in the wee hours of morning. Although the job paid well, the workers hated waking up in the middle of the night for a couple of months, and by the end of the season they were very sick. I wondered whether their sickness was some type of reaction to stuff in the field, or if it was due to losing sleep for 2 months. The field of circadian rhythm research was beginning to make headlines in top scientific journals as Jeffrey C. Hall, Michael Rosbash, and Michael W. Young (who together won the Nobel Prize in Physiology or Medicine in 2017) were publishing their groundbreaking work.

I soon left India for graduate school in Winnipeg, Manitoba, Canada. It was a profound shock on many levels, the least of which was moving from 98°F weather in India to Winnipeg, where 0°F temperatures during winter were not uncommon. The nights were so long in winter, and my brain was disoriented: Was it culture shock, temperature shock, or the lack of light? Almost half of my classmates in the immunology department were feeling quite low, and they called it the "winter blues." The effect of long Winnipeg nights on my circadian rhythm and mood rekindled my interest in the field. After just one winter, I managed to move to San Diego. That's where I put all of my life's questions and experiences into one field of research. I began to formally study circadian rhythms.

For the past 21 years, I've devoted my life to this research. As a graduate student at the Scripps Research Institute in La Jolla, California, I worked on understanding how plants measure time. The most exciting part was being in a lab that was at the forefront of the field. This was when we first discovered that there were clock genes in both plants and animals. Our work involved uncovering the mystery of how these clocks work. Every day was thrilling, almost like sitting in the front row of your favorite Broadway show every night. I was part of the team that discovered how specific plant clock genes work together to tell plants when to photosynthesize and absorb carbon dioxide for fuel, and when to sleep or repair themselves. One of the plant genes I discovered allowed us a better understanding of how the circadian clock, metabolism, and DNA repair may be connected.

In 2001, I was invited to do my postdoctoral research at the newly founded Genomics Institute of the Novartis Research Foundation (GNF), where I would be working on animal clocks. This premier institute squarely focused on using the newly described human and mouse genome to understand biology. I was there to solve mysteries in circadian biology.

My first breakthrough came in the first year. I was able to explain how our circadian rhythms adjust to different seasons or different types of light. My team discovered an elusive blue light sensor in the eye's retina that sends light signals to the brain clock to tell it when it is morning and when it is night. Having a handle on the light sensor helped us figure out how much light—of which color, for how long, and at what time of the day—we need to advance or delay our clock. That was a huge discovery because for almost 100 years, scientists had known that there was a light sensor in the eye, but they didn't have any idea where it was or what it did. This discovery was cited among the top ten breakthroughs of 2002 by the prestigious *Science* magazine, and it is the reason why your smartphone or tablet lets you change its background color from bright white to a dimmer orange a few hours before your scheduled sleep time.

It took us almost 8 years to determine how this light sensor worked, how it transfers information from the eye to the brain, and which brain regions receive that information in order to regulate sleep, depression, circadian rhythm, and pain. Even today, I am still trying to figure out the full extent to which light affects circadian rhythm and how modern lighting affects this process. Yet it's been very gratifying to see how our discovery went from simple observation to adoption, enabling more than a billion people to be aware of the impact of light on their health in just 15 years.

A second point of inquiry was to determine how our internal clocks transmit their timing information and how our organs read time and do different tasks at specific times. We started using very modern genomic technology to monitor which genes turn on and off at different times in different organs. This research began in 2002, and since then we have had another big breakthrough: the discovery that hundreds to thousands of genes in both the brain and liver turn on and off at specific times. We are still extending these experiments to different organs, tissues, brain centers, and glands. We are finding that almost every organ has its own clock, and in each organ genes turn on or off, affecting protein production levels at predictable times of the day.

After starting my own lab at the prestigious Salk Institute for Biological Studies, I continued my clock research in collaboration with outstanding colleagues. We now know that to have predictable circadian rhythms is to have healthy organs. Just like a mutation in the genetic code can lead to disease, living in opposition to the circadian code can push us toward disease. Over the past few years I have had the good fortune to work with some of the great minds in the fields of cardiovascular and metabolic diseases, and together we have found that animals that lack a normal clock are highly predisposed to these diseases. Slowly it became clear that a disrupted clock is the mother of all maladies, and, conversely, in most chronic diseases, clock function is compromised.

Finally, in 2009, these two areas of my research—light and time—came together. Extending the research of two prior studies, we created a simple experiment where we kept mice in a specific light-dark cycle.[1,2] Mice are usually nocturnal and eat at night. But in the experiment, we gave them food during the daytime and then watched to see what happened with their internal clocks. Surprisingly, we found that almost every liver gene that ever turns on and off within a 24-hour period completely ignored the light signal and instead were synced to when the mice ate and fasted. We also learned from this experiment that a daily eating–fasting cycle drives almost every rhythm in the liver. Instead of thinking that all timing information comes from the outside world through the eye's blue light sensor, we learned that just like the first light of the morning resets our brain clock, the first bite of the morning resets all other organ clocks.

Then, in 2012, we pushed the envelope even further. We wanted to see if disease was not only linked to diet but also to the breakdown of the circadian code. Thousands of research papers had shown that when mice are given free access to fatty and sugary foods, they become obese and diabetic within a few weeks. We compared one set of mice having free access to the fatty diet to a second group that had to eat all their food within an 8- to 12-hour period. What we found was startling: Mice that eat the same number of calories from the same foods within 12 hours or less every day are completely protected from obesity, diabetes, liver, and heart disease. More surprising, when we put sick mice on this scheduled feeding, we could reverse their disease without medication or change in diet.

Initially, the scientific community was skeptical of our discovery. The conventional wisdom was that what and how much we eat determined our health. But slowly, similar observations began to pour in from laboratories around the world, including from human studies. Now we know that in addition to what and how much we eat, when we eat matters.

Many important medical groups have taken note of our findings and have done their own literature review to find if timing of food intake matters. For instance, the National Institutes of Health, the American Heart Association, and the American Diabetes Association, among others, believe as I do that resetting the circadian clock is our next, best hope to prevent or to accelerate the cure of chronic diseases. In 2017, the American Heart Association released their first recommendation on meal timing and frequency in almost 70 years that corroborates our research, showing how eating patterns might be used as a way to prevent or reduce cardiovascular disease.[3]

This book, based on my research, is meant to give you the tools you can use to optimize your clock by making simple lifestyle changes. The stakes have never been higher. Today, almost one-third of all adults suffer from at least one chronic disease, such as obesity, diabetes, cardiovascular disease, hypertension, respiratory disease, asthma, or chronic inflammation. By the time of retirement, adults in the United States typically have two or more chronic diseases. And the truth about chronic disease is that there is rarely a cure. There aren't many people with diabetes who go completely back to normal. A person with cardiovascular disease rarely goes back to normal. We just have better ways to manage and live with these diseases.

That changes now. In this book I offer you very simple ideas and practices you can use on a daily basis that have been proven in vigorous laboratory research to prevent or delay the onset of disease.

Here's just one more thing you need to know about me: My science is supported by the U.S. government and thrives because of honest taxpayers and philanthropists like you. If this research can inspire a million people to make these small changes and delay one chronic disease by just one year, it can provide an estimated savings of at least $2 billion annually to the U.S. economy. This research is my gift to you because I feel so deeply indebted to this country. In 2001, I was a foreign national having just finished my PhD with an F-1 visa. I was very excited to continue my postdoc research at GNF and had just applied for an H-1B visa. Any foreign national knows the gut-wrenching anxiety of waiting for your work visa.

Then 9/11 happened. At about 5:00 p.m. on September 12, 2001, the human resource director of GNF walked toward my desk with a piece of paper in her hand. My worst fear came to my mind: that the government must have rejected my H-1B visa. But instead, I learned that it had been approved earlier that day. It was then I realized that this country, my new home, must be awesome, because on September 12, when I could not focus on my work in the lab, being completely overcome with the previous day's events, somebody on the East Coast actually went to work, looked at my application, and approved it. That was the day I decided to stay in this country forever and pay it forward. This is why I'm sharing my research with you, and I hope that you can benefit from it.

How This Book Works

Addressing your circadian clock is more than a diet. In fact, it's not a diet at all. It's a lifestyle. It begins with knowing when to eat and when to turn off the lights. Just paying attention to those small parts of your day will go a long way toward preventing and delaying disease.

As you'll learn, we are easily vulnerable to disrupting our circadian rhythm. All it takes is the slightest upset from an overnight flight, a poor night's sleep, illness, or a disruptive work schedule. *The Circadian Code* can be a powerful tool to manage your waking day, whether you are a parent or a child (especially a teenager), millennial or retired; regular workers, shift workers, working moms, and health enthusiasts can all benefit. If you are dealing with one or more chronic diseases, you need to read this book. No matter who you are, you'll learn when is the best time for you to eat, work, and exercise during the day, and how to manage the evening hours so that you can get the best, most restful sleep.

First and foremost, this book is about prevention, but you can also use this information to live better now. Part I focuses on identifying how the circadian clocks in the body work and why maintaining perfect timing is of utmost importance for both children and adults. The first step on the road to health is to recognize if you are in fact unwell, and this section includes a simple quiz to see how your health is currently affecting your rhythm. You will also start tracking your timing so you can see where adjustments need to be made.

Part II features complete instructions on how to best use your day to maximize your internal rhythms. You will learn exactly when (and what) to eat, but not how much. There is no calorie counting on this program, but I can say that if you do follow the guidelines I suggest, weight loss is almost inevitable. You'll learn when is the optimal time of day to work and be productive, as well as when is the best time to exercise. You'll also discover new techniques for getting a better night's sleep, as well as technology that can enhance and track your total experience.

As we age, disruptions to our circadian rhythm affect us more than when we are young. I believe that most of the diseases that affect us in adulthood can be traced back to circadian disruption. Part III addresses specific ailments and how they relate to our circadian rhythms. This section covers cancers and other immune-system issues, the components of metabolic syndrome (heart disease, obesity, and diabetes), and neurological health including depression, dementia, Parkinson's disease, and other neurodegenerative issues. You'll also learn how the gut's microbiome is influenced by your internal rhythms and how conditions like acid reflux, heartburn, and inflammatory bowel disease can be addressed.

I'm not a medical doctor, so I cannot prescribe medications. The scientist in me reminds me every day how little we really know about how the body works. But I can share with great certainty what I know about this powerful, primordial inescapable rhythm we have, including my best advice for optimizing your daily routines. Please share this information about daily habits that optimize our circadian rhythms with your doctor or other health practitioner so that he or she can make better decisions about treatment options or courses of action. With the tools given inside this book, it's very likely that you'll be able to get your health back on track.

PART I

The Circadian Clock

CHAPTER 1

We Are All Shift Workers

If you are a card-carrying shift worker who wakes up in the middle of the night to go to work, returns from work late at night, or stays awake all night, you know how it feels to be living against a primitive, primordial drive to sleep at night and stay awake during the day. But even if you're not, I'm sure you can remember a time when you were fighting against your internal clock. The truth is, we are all shift workers. There are times in life when we go through chronic sleep disruption, and for many, those habits linger. If you pull an all-nighter at school or work, stay up late studying for a test, have a bad night's sleep, travel across several time zones, stay awake late into the night to tend to a sick relative, or wake up a few times to feed and change a baby, then you too are a shift worker. A full-time job with long commutes combined with a regular home routine is like working two shifts and going to bed past midnight. Even one late night of partying can be just as disruptive as traveling from one time zone to another: That's why we call it *social jet lag*.

The statement that "we are all shift workers" isn't just an idea. Data points to this fact. For example, Professor Till Roenneberg, a researcher in Munich, surveyed more than 50,000 people in Europe and the United States and found that the majority of people either go to bed after midnight or wake up early with insufficient sleep.[1,2] Similarly, people also follow different bedtime schedules on weekdays and weekends. At the 2017 World Sleep Congress, Roenneberg presented his data showing that roughly 87 percent of adults have social jet lag and go to bed at least 2 hours later on the weekend.

About 6 years ago, my lab started monitoring the activity and sleep patterns of close to 200 college students, and we found the same pattern that Roenneberg has reported. So far, there's been only one person in the whole group who actually went to bed every day at the same time, within half an hour, including on weekends. There has been only one other student who went to bed before midnight for at least two days in a week.

We also monitor pregnant women and working moms with babies, and their patterns are also very erratic. In fact, their patterns are most similar to firefighters, who expect to be awoken a few times every night. For many women, the hardest part of motherhood is working against your clock to stay awake at night and trying to catch up on sleep at odd hours of the day. The only time new moms actually got good sleep, not surprisingly, was when they had some help beyond their spouse/partner, like in-laws or parents who could share some of the work at night.

Working mothers have the roughest time syncing their lives to a daily rhythm because their day is affected by everyone else in the home. Typically, working mothers wake up very early to get breakfast ready for the family, prepare the kids, pack the lunch bags and backpacks, get the kids to school or day care, and then get themselves to work. After dinner, they oversee homework, exercise, or work at home late into the night. As the week goes on, their circadian disruption becomes more severe. For instance, when my daughter was an infant, by Friday my wife would literally fall ill, and it would take her all weekend to recover.

No matter what the cause, we all know how it feels the day after a particularly rough night. You feel sleepy, yet you cannot sleep. Your stomach may feel upset, your muscles are weak, your mind is foggy, and you are certainly not in the mood to hit the gym. It's as if your body and mind are confused—half of your brain may be telling you that it is time to catch up on lost sleep, but the other half is insisting that it's daytime and you should not sleep. You may resolve to push on and reach for a strong cup of coffee or energy drink to stamp out the urge to sleep or try to get back into your regular routine as quickly as possible.

A brain on shift work cannot make rational decisions. According to a recent article in *Popular Science* magazine,[3] a single night shift has cognitive effects that can last a week. These lapses in memory or attention can also make us vulnerable to bad habits. A few days of reduced sleep can change our appetite, both for the kinds of foods we crave and how much we want to eat when we stay awake at night. Often, we are prone to eat more calorie-dense junk food late at night when our stomach is meant to rest and repair.

Living in the shift-work zone can also cause difficulty in getting to sleep. Some turn to alcohol or sleeping pills, both of which can trigger depression. But more important, they are addictive remedies that create bad habits that continue even when our lifestyle does not demand us to be awake at night.

And if it weren't bad enough that a shift-work lifestyle affects the way we feel the next day, our family members are in essence secondhand shift workers, as we may inadvertently disrupt their sleep as they wake up early or stay awake late to match our crazy schedules and keep us company. The effects on their health are equally troubling. For instance, in a 2013 analysis of published papers on the topic, researchers found that children of shift workers not only had more cognitive and behavioral problems as compared to children raised by non–shift workers, they also had a higher incidence of obesity.[4]

While a day or two of staying awake late into the night, or a couple of days after traveling through a few time zones, may be uncomfortable, repeatedly disrupting your circadian clock can have adverse health consequences, as every system in your body starts to malfunction. It makes the immune system so weak that germs and bugs that don't usually cause any trouble can upset your stomach or even cause flulike symptoms. It has been well documented that shift workers experience more health problems than non–shift workers, particularly gastrointestinal diseases, obesity, diabetes, and cardiovascular diseases.[5,6,7,8,9,10,11,12,13,14,15,16] Surprisingly, the number one cause of death and work disability for active-duty firefighters is not fire or accident—it is heart disease, which is now thought to be linked to a disruption of the circadian rhythm.[17, 18] In many studies, shift work increases the risk for certain types of cancer to such an extent that, in 2007, the World Health Organization's International Agency for Research on Cancer classified shift work as a potential carcinogen.[19]

If we are all shift workers, then we will all suffer. This is why we have to understand how our circadian clock works, and how to optimize our lifestyle to nurture the natural rhythm of the body.

What Happens When Circadian Rhythms Break Down?

ADHD
Autism
SAD
Anxiety
Panic attack
Depression
Compromised learning
Nocturnal epilepsy
Bipolar syndrome
ICU Delirium
Migraine
PTSD
Seizure
Mania
Psychosis
Multiple Sclerosis
Huntington Disease
Alzheimer's Disease
Parkinson's Disease
Bacterial Infection
Sleeping sickness
Malaria
Arthritis
Asthma
Allergy
Lymphoma

Polycystic ovarian syndrome
Irregular menstrual cycle
Post-partum depression
Inability to conceive
Morning sickness
Miscarriages

Insomnia
Prader-Willie syndrome
Smith-Magenis syndrome
Obstructive Sleep Apnea
Delayed sleep phase syndrome
Non-24-hour sleep-wake syndrome
Familial advance sleep phase syndrome

Leaky gut
Indigestion
Heart burn
Stomach pain
Crohn's disease
Ulcerative colitis
Inflammatory bowel syndrome
Inflammatory bowel disease
Metabolic syndrome
Weight gain/Obesity
Childhood obesity
Type 2 Diabetes
Prediabetes
Stroke
Dyslipidemia
Hypertension
Heart Arrhythmia
Chronic Kidney Disease
Fatty Liver Disease (NAFLD)
Steatohepatitis (NASH)
Ovarian cancer
Breast cancer
Liver Fibrosis
Colon cancer
Liver cancer
Lung cancer

Diseases linked to circadian disruption

Which Kind of Shift Worker Are You?

A person who stays awake for more than 3 hours between 10:00 p.m. and 5:00 a.m. for more than 50 days in a year fits the official European definition of a shift worker. Yet I believe we are all shift workers simply due to the way we live our lives. Which kind of shift work do you experience?

- *Traditional shift work:* Roughly 20 to 25 percent of the nonmilitary workforce in any developing or developed country is involved in shift work. This includes emergency responders (firefighters, emergency dispatchers); police; workers in health services (nurses, doctors), manufacturing, construction, utility services, air transportation (pilots, flight attendants, ground staff), ground transportation, and food services; custodial staff; and call center customer support workers.

- *Shift-work-like lifestyle:* This includes high school and college students, musicians, performing artists, new mothers, in-home caregivers, and spouses of shift workers.

- *Jobs in the gig economy:* This includes part-time drivers for ride-share services and food delivery services, flexible workers, and freelancers.

- *Jet lag:* This occurs when you travel across two or more time zones within a day. Nearly 8 million air travelers take to the air each day,[20] and half of them travel over at least two time zones.

- *Social jet lag:* This occurs when someone sleeps late and wakes up at least 2 hours later on the weekends. More than 50 percent of the population in modern society experiences social jet lag.

- *Digital jet lag:* This happens when you chat with friends or colleagues that are several time zones away over social networks or digital devices and as a result have to stay awake for more than 3 hours between 10:00 p.m. and 5:00 a.m.

- *Seasonal circadian disruption:* Millions of people living in extreme north and south latitudes (residents of northern Canada, Sweden, Norway, and southern Chile, for example) experience less than 8 hours of daylight during winter and more than 16 hours of daylight in summer. These extreme exposures disrupt their circadian rhythm.

Circadian Rhythms Are Real

We used to believe that our day-night cycles were only guided by the external world: The light in the morning would wake us up, and seeing the moon was our cue to go to sleep. Many scientists discounted the entire field of circadian biology, even until the mid-1970s. While it was known as far back as 1700 that there was an internal clock in plants, the idea that animals and humans were internally driven rather than externally motivated was hard to prove. The common wisdom was that humans, a more evolved species, must be driven by outside or environmental factors beyond the sun and moon.

The plant experiments were easy enough: A plant placed in a dark basement will still move its leaves up and down in a particular rhythm each day.[21] Many plants move their leaves up during the day to capture more energy from sunlight. At night, their leaves can drop, because it would be a waste of energy to keep the leaves raised. Similarly, many flowers only bloom during the day, when pollinating bees and birds are flying around, yet some, like the jasmine tree near my grandparents' home, bloom at night: These plants depend on wind, not other animals, for pollination.

The next set of studies was exponentially more difficult, and scientists started with insects, birds, and then animals. They researched the timing of larvae turning into fruit flies, which is circadian because it only happens in the morning, when there is less wind and more humidity. They studied the migration patterns of birds and the waking patterns of other animals. Laboratory mice were also studied under a controlled environment.[22] When they were put under constant darkness without any outside timing cues, they also woke up and went to sleep with clockwork precision, every 23 hours 45 minutes. Similarly, the circadian clocks of many plants and fungi are close to but not exactly 24 hours.

It was almost impossible to investigate if humans had these same internal clocks because there was no easy way to remove all of the external timing cues of every connection to the outside world. However, in the 1950s, researchers had an idea: They created a simple telephone that could connect a volunteer to only one other person. The volunteer went deep into a cave far in the Andes Mountains. All he brought with him was enough food, candles, and reading materials to keep him occupied for weeks. Each time he felt sleepy enough to go to bed, he would call his partner on the other side of the phone, who would record the time. He would make the same call when he awoke. The study showed that his sleep-wake cycle continued with clockwork precision for several weeks in the cave. However, the volunteer went to bed a little later every day, implying his clock was slightly longer than 24 hours. In fact, he was going to bed and waking up within a cycle that covered exactly 24 hours 15 minutes. His cycle was so predictable that it could only be guided by an internal clock.[23]

The fact that the circadian rhythm is not exactly 24 hours is not surprising, as the timing of sunrise to sunrise in most of the world is not precisely 24 hours. Because our planet has a tilt to its vertical axis, as it travels around the sun there are times during the year when either the Northern or Southern Hemisphere stays facing the sun longer. As the daytime slowly gets longer or shorter over the course of the year, the sunrise and sunset times change. At the equator, the change is very small, but if you live in Boston, Stockholm, or Melbourne, the sunrise time from one day to the next can change as much as a few minutes. When daylight lengthens as we approach summer, our internal clock wakes us up at a slightly earlier time in the morning, just when the sun comes up. When we fly from one time zone to another, our sleep-wake cycle slowly adjusts to the new time zone. These examples are just a few that explain why we have an internal clock, and how its mechanism to adjust itself connects to the change in sunrise time or day length. Once this was determined, scientists surmised that circadian rhythms are connected to, or can be timed to, light.

The Rhythm of Daily Life

Scientists like me continue to examine the daily rhythms in adult human physiology, metabolism, even cognition, and we have found that almost every aspect of our daily life is rhythmic. Although humans don't flower, or migrate over long distances, we do have circadian clocks that time almost every aspect of our daily health to the right moment of the day or night. In fact, our body is programmed to go through specific rhythms every day. Interestingly, your evening activities have a large effect on your circadian rhythm. The changes you'll make from reading this book that will be most profound will be created by monitoring your life from 6:00 p.m. to midnight.

Even before we wake up in the morning, our internal clock prepares our body for waking up. It begins to shut down the production of the sleep hormone melatonin from our pineal gland. Our breathing becomes slightly faster and our heartbeat picks up a few beats per minute as our blood pressure rises slightly. Our core body temperature notches up half a degree even before we open our eyes.

Our entire sense of health is guided by our daily rhythms. In the morning, being in good health means waking up feeling rested and re-freshed from having a good night's sleep, having a healthy bowel move-ment to eliminate the toxins we collected at night, and feeling alert, light, and hungry for breakfast. Shortly after we open our eyes, the adre-nal glands produce more of the stress hormone cortisol to help us rush through our morning routine. The pancreas becomes primed to release insulin to handle breakfast.

After a good night's sleep and nourishment from breakfast, the brain is primed for learning and problem solving in the first half of the day. In the afternoon, we feel healthy if we have accomplished enough work to feel satisfied with our efforts. (When you did not have a good night's sleep the night before, you may have an overwhelming feeling that you're wasting the day.) As the day goes by, muscle tone peaks toward the end

The Body's Daily Rhythms

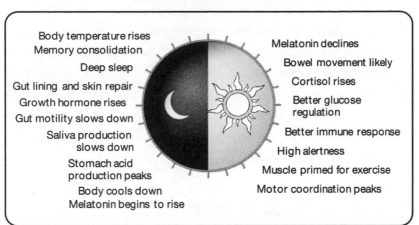

Body temperature rises
Memory consolidation
Deep sleep
Gut lining and skin repair
Growth hormone rises
Gut motility slows down
Saliva production
slows down
Stomach acid
production peaks
Body cools down
Melatonin begins to rise

Melatonin declines
Bowel movement likely
Cortisol rises
Better glucose
regulation
Better immune response
High alertness
Muscle primed for exercise
Motor coordination peaks

Many of our body's functions peak at certain times of the day or night.
These rhythms are thought to be regulated by our circadian clocks.
They will continue on their normal schedule for just a few days if we are
completely removed from the natural cycle of day and night.

of the day. And as the sun sets and evening rolls in, our body temperature begins to drop, the production of the sleep hormone melatonin begins to rise, and the body prepares to sleep.

In the evening, being in good health means winding down, feeling tired, and falling into a deep sleep without much effort. Sleep is not a default mode during which the brain just shuts down. In fact, the brain is very busy as we sleep. It is consolidating memories based on the sensory information we took in during the day by backing up this information as it creates new synapses, or connections, between different neurons. The brain also produces quite a few hormones at night. The sleep hormone melatonin is produced in the brain's pineal gland. Human growth hormone is also produced when we sleep.[24] In fact, people who have insufficient sleep produce less growth hormone. This is extremely important for children because a lack of sleep can reduce the amount of this important hormone and can hinder growth.

At night, the brain also detoxifies. During the daytime, brain cells absorb and process nutrients, creating unwanted toxic by-products. These toxins are cleaned up when we sleep, and new brain cells are created through the process of neurogenesis. In this way, our brain is like an office: When you come into the office in the morning, you don't think that anyone was working overnight, but actually a lot of things were happening. The trash was taken out, and the repair people might have come in to upgrade the servers or replace the light bulbs. All this work has to go on so that you can come in and start a fresh new day.

We Need Strong Circadian Rhythms

Circadian rhythms optimize biological functions. Every function in the body has a specific time because the body cannot accomplish all it needs to do at once. Watching newborn babies gives us a better appreciation of why we need circadian rhythms. We've learned from observing the developmental patterns of newborns that babies come into this world without a very functional circadian clock: Their rhythms are apparent, but not robust. For instance, babies try to sleep, but in the middle of the night they become hungry or they poop, and either of those biological needs is strong enough to wake them up. Then they cry because they're hungry or messy and sleepy at the same time. Everything is chaotic. However, as their circadian rhythm strengthens, at around 5 to 8 months, it can exert more control over their body functions. The first thing that happens is that they can experience uninterrupted sleep for several hours. Their digestion slows down so that they do not need to be fed at night, and they can hold their bowels until the morning because the hormone levels that promote bowel movement are suppressed during sleep. Every day, the rhythms strengthen and become more entrenched.

As babies grow into toddlers, family life begins to assign times to bodily activities. We have a prescribed time for breakfast, lunch, and dinner. At the same time, the light sensors in our eyes are programmed to notice changes in the timing of morning light and adjust our internal clock slightly by a few seconds or minutes every day. This "light entrainment," or syncing the internal clock to the natural day-night cycle, enabled our ancestors to wake up at dawn, no matter the season.

The circadian clock is the internal timing system that interacts with the timing of light and food to produce our daily rhythms. Our job is to maintain the clock so we can live with optimal health. As you'll learn, the best way to do this is to live in accordance with the circadian clock, rather than push against it. First, let's discover the role that light plays.

A Brief History of Harnessing Light

All of human history can be summed up as our attempt to beat the clock, as primordial rhythms evolved to predict and adapt to the environment. In order to understand how light affects behavior, we need to focus our attention on evolutionary biology, which traces our heritage back roughly 2 million years and connects to the adaptive mechanisms we've developed to survive in any environment. We know our evolution is relevant today because our physiology—the way we are meant to function—is largely the same today as it was 2 million years ago. We are still meant to sleep at night and work and eat during the day, on a cycle that is programmed by our internal clock.

We know that modern humans largely evolved near the equator and their daily activity was directed by the sun and influenced by a very strong corresponding circadian rhythm. Primitive men and women had to wake up before the sun came up if they wanted to be successful hunters: Their strategy was to wait for their meal to stroll by a water hole. If they could not hunt, they would have plenty of time to explore and collect berries and fruits. Finding and eating food took a very long time, especially if they also had to stay away from predators.

They also had to have enough muscle tone in the late afternoon to run back the few miles they traveled away from the cave or shelter in their search for food. Anthropologists assume that early humans ate their last meal just around twilight, leaving them plenty of time to find a safe place to sleep before nightfall. At night, they rested for 12 to 15 hours, a big chunk of which was spent sleeping. This nighttime fast must have helped cleanse the gut so that in the morning they would be light and ready to go back to hunt for more food.

Humans have the unique ability to voluntarily switch their lifestyle from day to night, staying awake through the night when necessary, changing and challenging our circadian rhythm. We are unusually equipped to adjust our circadian rhythm because large animals posed a threat, so we had to evolve a way to stay awake, even for a few minutes, at night, in the dark. Individuals would take turns watching over the rest of the community while the others slept: These were the first shift workers.

Winning the night was the ticket not only to survival but to prosperity and wealth. Many hunters learned to prefer to hunt at night. These shift workers became an essential part of human society. Over time, explorers and conquerors that could navigate their way through the night and mount surprise attacks against their enemy became both prosperous and wealthy by expanding their territory and acquiring new farmland, minerals, precious stones, and natural resources.

Fire was the first tool humans used to work against their clock. The ability to start and control fire gave humans two advantages: First was the light itself, which allowed us to stay awake for a few extra hours, and through the night if need be. The evening flickering light from burning embers was dim—just enough to allow early people to find their way, deter large carnivores, and offer overnight warmth. Second, fire became a powerful weapon. For thousands of years the only weapon we had was fire. Even now, most of our weapons are still based on fire.

Life around the fire pit also fueled the rise of human civilization. Fire was essential for cooking food and boiling water, extending the types of food that could be eaten. Cooking tenderizes food, breaks down strong flavors and makes food more palatable, and kills pathogens and makes it safer to eat.[25] The cooking process also makes food more digestible, so we can squeeze more calories from the same ingredients. This is why eating raw food can be a weight-loss strategy but cooking then eating the same food does not affect weight loss as much.[26] Cooking also reduced the time we spent hunting for food because we could extract twice as much energy from the same food. At the same time, we had more options to choose from: We could now eat many foods that could not be digested in their raw form.

As fire offered warmth on cold nights, it allowed early humans to move away from the equator and venture to higher latitudes in northern Europe, Asia, and North America. Humans reached the northernmost latitudes relatively recently, only 30,000 to 40,000 years ago. In summertime, the long day, sometimes stretching to 20-plus hours of light, was not too difficult to adjust to because summers were not that hot and humans could catch enough sleep in dark caves or huts. But the long winter night with very little daylight would have certainly confused the brain. Even today, many people cannot adjust to long, dark winter nights in higher latitudes and develop seasonal affective disorder, or seasonal depression. The rates of both depression and suicide attempts in these areas increase in the winter, which we now understand to be connected to circadian disruption: It is as if people who suffer from seasonal depression are stuck working the night shift for several weeks or months.

No matter where early humans lived, fire also had a very special effect on the evening life. While the men's day was spent hunting, the women and children remained near the home, tending to domesticated animals or drying and processing food for rainy or winter days. The evening fire pit brought everyone back together, creating a special time for families to entertain themselves, relax, and unwind. People would share stories, plan for the future, think imaginatively, and develop new ideas in science, culture, and crafts. Evening fireside talk is the cradle of the art, culture, science, and philosophy of the things that make us human.[27] This evening social life around light is deeply ingrained in our daily living.

But this evening fireside time was restricted to an hour or two, because maintaining the fire was difficult, and it would later come to be relatively expensive. Even during the early eras of industrialization, fire and access to light was rare. As humans moved to whale oil, beeswax, and tallow as better fuel sources, a distinction was often made between fire for cooking or heating and fire for light. Using these fuels for light was too expensive for average people. In today's dollar value, it would cost $1,000 to $1,500 to light up an average 19th-century home for a few hours every evening.[28] As bright light in the evening was rare, most people felt sleepy and went to bed just a couple of hours after sunset through the 19th century. Today there are indigenous populations in Africa, South America, Australia, and India who live an agrarian or hunter-gatherer lifestyle similar to the lifestyles of 2 to 3 centuries ago. In these communities without much access to electricity, the people go to bed early and wake up around dawn.[29,30,31]

At the turn of the 20th century, electricity and electric light spread throughout the Western world, yet there was still not much reason to stay awake and do much at night. Gas and electric stoves continued to uncouple heating from traditional wood fires, bringing the kitchen from outside to the center of the modern home and making it safe to cook food at whatever time we wanted. Food processing and preservation techniques, along with refrigeration, made it possible to have access to food all the time. And that's when the trouble really began.

Early industrialization increased food production as well as mining and manufacturing, now requiring less physical labor both on the job and at home. Increased production soon exceeded local consumption, which led to the development of infrastructure—highways and trains, buildings and storage—further reducing demands on human physical activity. Maintenance and creation of this modern infrastructure also required a new breed of workers who would stay awake and work at night. Today, nearly 20 to 25 percent of full-time workers in industrialized societies are shift workers.

Mechanization of agriculture in the early 20th century also increased crop yield, while plant breeders unknowingly selected plants that had naturally tweaked their circadian clocks. These "mutant" crops did not need to correctly calculate day length to figure out whether it was summer or winter. Instead of being restricted to flowering on long summer days or short winter days, these crops could flower in any season or, like tomatoes, in a greenhouse, so farmers could raise two to three crop yields from the same land every year, further boosting production.

As food production became mechanized, it untethered workers from spending all day outdoors. Meanwhile, electric lighting became increasingly affordable. Fast-forward to the mid-20th century. After World War II, with all of these industrial systems in place, almost everyone in industrialized nations started experiencing circadian disruption. Sleeping less also meant increasing the time we spent awake under bright lights, especially at night when the brain does not expect to be stimulated by light. And when we were awake during the day, many of us stayed indoors and didn't get enough exposure to bright sunlight. Both of these scenarios confuse the brain clock.

Telephones, radio, and television began entertaining us late into the night. The computer has taken the local evening fireside talk and completely transformed it into a real, yet virtual, global 24-7 chat session in which you can discuss any topic with anyone, anywhere in the world. And with a 24-hour news and entertainment cycle and billions of computer devices in use across the globe, who can afford not to be connected?

Yet while all of these advances are supposed to update the previous technology and make our lives better, they increasingly disrupt our body's clock. Our circadian rhythm continues to be confused by bright light in the evening and limited access to natural light during the day. We simply have not evolved enough to sync our internal clock with the realities of the modern world in which we live, and consequently, we are all struggling like our northernmost ancestors, or even our current Nordic cousins. Regardless of whether we are real shift workers or just live a shift-worker lifestyle, constant exposure to light at night causes circadian disruption that suppresses sleep and leaves us hungry.

Light for Health Is Not The Same as Light for Sight

We cannot travel back in time to the Middle Ages to take advantage of a long, dark night, but if we know how light affects our clock, perhaps we

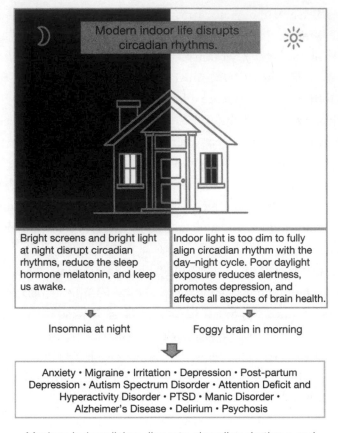

Modern indoor living disrupts circadian rhythms and predisposes us to a range of brain diseases.

can master light to master our health. When I started graduate school, I had lots of questions: I wanted to know exactly how light influences the internal circadian clock. Why does staring at a computer screen at night keep us awake, while in the morning our brain clearly needs a lot more light to stay alert? Is there a color of light that is more potent in affecting our clock?

If we could figure out how brightness and color of light affect our clock at different times of the day, we could control the use of light to boost our health. Although you might know that the skin's exposure to bright sunlight is necessary for making vitamin D, it has nothing to do with our clock. All influence of light on our clock goes through our eyes. So, let's discuss how our eyes work.

The human eye works like a camera. It contains millions of rod and cone cells that capture the details of an image in fine resolution and send this information through long, wirelike nerve cells to the brain. The retina, the light-sensitive tissue lining the back of our eye, contains several million rod and cone light sensors. Light rays are focused onto the retina through our cornea, pupil, and lens. The retina converts the light rays into impulses that travel through the optic nerve to our brain, where they are interpreted as the images we see. We lose the ability to see when these rod and cone cells die, as in some congenital cases of blindness.

Yet blind people have circadian clocks that are still influenced by light. Surprisingly, many blind people can still "sense" light. As they walk into sunlight, many report that they can feel some brightness filling their eyes, and their pupils actually get smaller under bright light and larger when they walk back indoors. These blind people and some blind animals can align their sleep and wake-up time to seasonal changes in day length.

This phenomenon was realized in the early 20th century, and for nearly 80 years, most scientists believed that the reason was that blind people might still have working rod and cone cells that would provide them with the sense of light. However, very careful experiments done in the 1990s showed that instead, there was an elusive light sensor in the eyes that we didn't know about.[32,33,34] In 2002, three independent research groups, including mine, discovered a light-sensing protein present outside the rod and cone cells that is, in fact, the light sensor that entrains the daily sleep-wake cycle to light.[35,36,37,38] This light-sensing protein is called *melanopsin*.[39] Of the 100,000 retinal neural cells that transfer all light information to the brain, only 5,000 contain melanopsin. Rod and cone cells can also entrain the circadian clock, but only in the absence of melanopsin, and when they do, they are not as efficient. This is why blind people who lose the rod and cone cells but still have intact retinal cells can still sense light. But these cells are so sparse that they are not sufficient to produce an image of the outside world.

To understand how this light sensor works, in our experiment we used mice that either lacked the melanopsin gene or the melanopsin cells, even though their eyes were otherwise perfectly normal—they could see fine and find their way around. When the gene is bred out of the mice, the cells stay alive, but when the cells are removed, the genetic expression is ended as well. When the melanopsin gene is removed, light information can still trickle into the mouse's brain through melanopsin cells. But when the cells are gone, all connection between the eye and the brain's circadian clock system is gone.

Normal mice usually wake up in the evening (they are nocturnal) and sleep during the day. But mice that don't have melanopsin cells cannot sense the light and darkness. Yet when these mice were put under constant darkness, they maintained a normal circadian clock—they went to sleep and woke up just like a normal mouse does, the cycle repeating every 23 hours 45 minutes. However, the melanopsin-free mice had greater difficulty adjusting to the small time change that occurs during any given week. While normal mice could readjust their sleep and wake-up time to the light-dark cycle within a week, the mice that didn't have the melanopsin gene took an entire month or longer to adjust. What's more, normal mice—like deer—freeze when they see bright light at night. But the mice that didn't have melanopsin didn't freeze under bright light at night: They continued running around. Finally, light at night did not affect the melatonin-production system in the mice that lacked both melanopsin genes and cells.

Since mice and humans share most of the same genes, including melanopsin, experiments with mice have immediate implications for human circadian rhythm. They suggest that melanopsin can impact the human circadian clock, our sleep cycles, and melatonin production. Our next question was targeted to better understand what type of light can be most or least effective for activating melanopsin so that we can have the right type of light at the right time to optimize our clock.

Visible light includes all the colors of the rainbow. Each color has a different wavelength. Red has the longest wavelength, and violet has the shortest. When all the waves are seen together, they make white light, or sunlight. Different colors within that white light activate three different types of opsin proteins (red, green, and blue), which in turn identify these colors individually and collectively (as white light). The melanopsin protein is most sensitive to blue light waves and is less sensitive to red light. When melanopsin is activated by registering blue light, it sends a signal to the brain that any light is present, and the brain responds by thinking it is daytime, regardless of what time it really is. If you're walking in the grocery store at night, your melanopsin is registering the overhead light and your brain thinks that it's daytime and you should be awake.

Imagine that you have two light bulbs with identical brightness—one is a blue light and the other one is an orange light. In the middle of the night as you turn on the orange light, the light fires up opsins in the green cones (green cone opsin can sense some orange light as orange is close to green in the rainbow) and your brain recognizes what is in the room. If you turn on the blue light, your blue cones will fire up and you can see the same objects in the room. However, melanopsin cells barely fire up under orange light and will tell the brain it is night, while the blue light will register as daylight. So, if you spend an hour under orange light, your circadian clock may not be disturbed much, but spending an hour under blue light will make your clock reset as if it is morning.

As seasons change and day length changes, our circadian rhythms adjust to the change in sunrise and sunset times. For a long time, we did not have a clear idea of how these circadian rhythms reset to new timing of sunrise or sunset, or how circadian rhythms are influenced by light. But our research showed that these same blue light sensors reset the brain clock when day length changes with each season or when we travel over different time zones. They also directly or indirectly connect to brain regions that control depression, alertness, sleep, production of the sleep hormone melatonin, and even to the brain center that regulates migraine pain or headache.

Melanopsin has another peculiar property: It takes a lot of light to activate it. For example, if you open your eyes for a few seconds in a dimly lit room, your rod and cone cells can take in an image of the room, but your melanopsin cells will react as if it was too dark to see.

These discoveries helped us begin to understand how light affects health. Our modern lifestyle, in which we spend most of our time indoors looking at bright screens and turn on bright lights at night, activates melanopsin at the wrong times of day and night, which then disrupts our circadian rhythms and reduces the production of the sleep hormone melatonin; as a result, we cannot get restorative sleep. When we wake up the next day and spend most of the day indoors, the dim indoor light cannot fully activate melanopsin, which means that we cannot align our circadian clock to the day-night cycle, making us feel sleepy and less alert. After a few days or weeks, we get into depression and anxiety.

Now that we have a better understanding of the quality, quantity, and duration of light that can shape or break our health, we can begin to imagine how simple changes to our light bulbs, computer screens, or eye glasses can go a long way to restoring or improving our health.

CHAPTER 2

How Circadian Rhythms Work:
Timing Is Everything

The second part of my research uncovered new information about our internal clock. All living things on our planet undergo an unescapable and predictable daily change in their environment: Day becomes night. It does not matter whether they are living in the desert, mountains, tropical forests, or whether they lived a billion years ago or are alive today. To cope with this predictable daily change in light and dark, almost every living organism has developed an internal timing system, or circadian clock.

Every living organism spends its 24-hour day:

- getting energy (food)

- optimizing energy use by using some for maintaining daily functioning and storing the rest for later use

- protecting itself from harmful agents and predators

- repairing itself or growing

- reproducing

All of these functions are guided by a circadian clock, which optimizes every organism's ability to carry out these tasks by assigning each of these essential aspects of life to an optimum time of the day or night.

Plants follow a roughly 24-hour circadian clock that predicts sunrise and sunset so they can optimally harvest sunlight and carbon dioxide to make food. The clock provides the rhythm; plants know to raise their leaves up an hour or two before sunrise and activate a number of genes so that they can begin to harness light from the sun's first rays. At the end of the day, plants shut down their light-harvesting machinery an hour or two before the sun goes down so that no effort is wasted in running the food-making factory when there is no light. Finally, their leaves droop around evening as if they are ready to go to sleep.

Plants also have daily rhythms that instruct them when to bloom, either by season or at certain times of the day or night. This plant rhythm is synced to the rhythm of pollinating bees and insects that feed on the plants' flowers. Large herbivores, like cows or camels, graze on plants during the day, and small rodents feast on fruits and vegetables at night to avoid their predators. In other words, they use their circadian clock to wake up, be active, and eat when it is safest. Even the bread mold *Neurospora* that grows on other food has a clock that instructs it to grow and make more spores on a daily 24-hour rhythm. Its spore-making function is timed to the appropriate time of the day that promotes optimal spore dispersion by the wind.

As you learned in the previous chapter, this exquisite timing might seem at first to be controlled by light. However, it was the exploration of genetics that showed researchers like me exactly how circadian clocks work. We learned that while circadian rhythms are influenced by light, the timing they follow is controlled internally, by genes.

The Genetics of the Circadian Clock

The human body is made up of millions of cells that are specialized based on location: There are cells that make up every body part, from your toes to your brain. Yet each of these millions of specialized cells contains the same *genome,* which is all of our hereditary information that we received from our parents. This information is encoded as our DNA, and the individual segments carrying this genetic information are called *genes.* Some genes correspond to visible traits, such as eye color. Others are related to biological traits, such as blood type, risk for specific diseases, as well as the thousands of biochemical processes, including our circadian clock.

These processes are then carried out by different types of proteins. Some proteins are enzymes that operate like construction tools (drill, hammer, chisel, etc.). Inside every cell, enzymes perform many tasks, such as making cholesterol and breaking up fat. Other proteins are structural; they are the building blocks of cells, like the parts of your home (walls, doors, etc.). Some tiny proteins are actually hormones (although not all hormones are small proteins), chemical messengers that control organ function. Some proteins last for a long time, while others are short-lived.

The health of our organs, and whether we have a particular illness, depends on which genes we have and how they are expressed: whether a specific gene is turned on or off, or if it is a normal gene or a mutant. For instance, have you ever noticed that some people can eat whatever they want, while others complain that certain foods, often dairy products, cause digestive discomfort resulting in gas, bloating, or constipation. Those that suffer actually have a mutation in the gene that helps break down and absorb the nutrients from milk.

By comparing mutant genes with normal ones, we can learn a lot about how genes are supposed to work, and the consequences of an abnormality. In the circadian field, scientists were first able to understand how our clock works by looking for mutant organisms whose clocks ran either too slow or too fast. In 1971, California Institute of Technology fruit fly geneticist Professor Seymour Benzer and graduate student Ron Konopka took thousands of fruit flies and studied them in isolation under constant darkness. Young flies are usually active at dawn and dusk, take a daytime siesta, and sleep at night. The fruit flies maintain these roughly 24-hour rhythms even under constant darkness. Benzer and Konopka built some truly ingenious tools to monitor when baby fruit flies went to sleep and woke up, even under complete darkness. After screening thousands of fruit flies, they found three types of mutants: flies that went to sleep early, late, or in no particular pattern.[1] They also found that the offspring of the mutant fruit flies inherited or maintained the same abnormal circadian clocks: That was the genetic component. The same mutation also changed the timing of when fruit flies hatched, which suggested that fruit flies have only one clock. Benzer and Konopka named this the *Period gene,* or *Per gene* for short.

The process of scientific inquiry is very much like solving a crime. From a few clues you can come up with the profile of a criminal, but it can take months or years to find the suspect and prove the crime. It took almost 13 years and two independent groups of scientists to figure out what the Per gene in fruit flies actually looked like. It took a few more years to figure out how the gene creates a clock.

Now we know that inside every cell, the Per gene sends instructions to create a protein that builds up slowly and then breaks down every 24 hours. This is true for every organism: There are three genes that control the clock in pond scum, and more than a dozen in animals and humans. Here's how it works: Let's imagine that a protein is an ice cube that's made in your freezer. The Per gene is the ice-making machine in the freezer and controls the exact amount of ice cubes that will be made. The freezer makes ice one cube at a time and drops each cube into a bin under the ice maker. After a couple dozen ice cubes build up in the bin, the bin becomes heavy enough and the machine turns itself off and stops making ice (likewise, the Per gene turns off once enough PER protein is made).

Every day, we take out all the ice cubes and make smoothies for the family. Then we put the bin back, and the ice-making machine restarts and continues making ice cubes until the bin is full. And because the machine's "Per gene" won't change, the number of ice cubes made every day is always the same, and the amount of time it takes for the machine to make the ice and for us to empty the ice bin is always the same. That time period is considered to be one cycle. If that cycle takes 24 hours to complete, then it is considered a circadian clock.

Now, if every ice-making machine worked perfectly all the time, we would all have identical rhythms every day. The problem is that how you take care of your ice maker affects its function. If you only take out a couple of ice cubes a day, the process of making an entire bin of ice cubes will take less time to complete. Similarly, at night, while the ice machine is making fresh ice cubes to fill up the bin, if you empty the bin again for making margaritas late at night, there will not be enough hours left for the machine to fill up the ice bin by morning. This is how you break the circadian rhythms when you stay awake under bright light or sleep late into the day.

A second problem occurs if you have a malfunctioning machine to begin with: That's a mutation. If the ice machine's "Per gene" is mutated, it may make ice too fast or too slow. The sensor that tells the machine to turn off could be faulty so that the machine stops making ice even if the bin is half-full or may continue to overflow the bin even if it is full. The faulty machine affects how long it takes to make each batch of ice cubes and use them completely every day.

Every Organ Has Its Own Clock

Scientists almost took it for granted that there was only one clock that controlled the entire body, and they assumed that clock resided in the brain, until one experiment by a PhD student smashed this supposition. Jeff Plautz, who was just a few years ahead of me in graduate school, took fruit flies and fused their Per genes with a glow-in-the dark fluorescent tag. These flies, with access to enough food and water, would glow green and dim down with a 24-hour rhythm, even when put in a completely dark room. One day, Plautz was cleaning up in his lab, and he chopped up a few live fruit flies and used the fly bits—wings, antennae, mouths, legs, abdomens, etc.—for a different experiment. He had heard that even after chopping up a fly, the individual organs would stay alive for a few days. He left for a Las Vegas vacation and returned after a week. When he walked back into his darkroom lab, he noticed that the antennae, legs, wings, and abdomens that had been completely separated from the heads of the flies were still glowing in perfect rhythm, just like a whole fly would do. The organs did not have to be attached to the body to glow/dim with a 24-hour rhythm. This experiment proved that every organ in an animal has its own clock, and these clocks don't need instructions from the brain in order to function. Plautz's discovery was named a top-ten breakthrough of 1997 by *Science* magazine.

Imagine the human body is like a house, and every organ is a different room, with a different clock. The clock in the bedroom tells you when to go to sleep and wake up, the clock in the home office tells you when you should work, the clock in kitchen tells you when to eat, and the clock in the bathroom tells you . . . You get the idea. Today we know that the clock in the gut times when to produce gut hormones for hunger or satiety, produce digestive juice to digest food, absorb nutrition, nudge the gut microbiome to do its job, and move waste out of the colon. The clock in the pancreas times when to produce more insulin and when to slow down. Similarly, clocks in muscle, liver, and the fatty tissue we accumulate do the respective job to tune the organ's function.

I took my research one step further outside the circadian clock genes and asked: How do clocks regulate the sleep tracker in the brain compared to how they control metabolism in the liver? While other researchers were focused on how a dozen clock genes turn on and off at different times of the day or night in the brain or in the liver, I wanted my team to cast a very wide net and test which of the more than 20,000 genes in our genome turn on and off at different times in different organs. We started a study in 2002,[2] using very modern genomic technology. And through that research, which is still continuing with increasing sophistication, we found that in every organ, thousands of genes turn on and off at different times in a synchronized fashion.

Every gene in our genome has a circadian cycle. However, they don't cycle at the same times, and some cycle only in one organ. This means that for every tissue there is a hidden time code to our genome. For instance, even though every single cell in our body contains a full genome, we found during the same 2002 study that up to 20 percent of all genes can be either turned on or off at different times of the day: Remember, we can't have all biological functions happening at the same time. What is more interesting is that the 20 percent of genes that are turned off for a specific time in the brain are not the same genes that are turned off in the liver or the heart or muscles. Having a detailed knowledge of the action of genes and their timing has given us a clear understanding of how circadian rhythm optimizes cell function.

Now let's take a look at what cellular activity occurs in a cyclic manner:

- The nutrient- or energy-sensing pathways—cell's hunger and satiety pathways—are circadian. Just like our whole body feels hungry when it runs low on readily available energy and gets satiated after we eat, or does not feel too hungry at night, every cell in every organ has such a mechanism that makes the cell hungry and opens the door to let nutrients flow in during the day; and when the cell has enough energy, it closes the door so that it does not get overstuffed.

- The energy metabolism pathway is circadian, affecting cellular function and metabolism of all key nutrients. The use and storage of carbohydrates, fat, or protein is not a continuous process. When sugar is absorbed from the blood and converted to fat or glycogen for future use, the body's fat-breakdown function is shut down. Only after the sugar is depleted does fat breakdown resume.

- Cellular maintenance mechanisms are circadian. Every chemical reaction, particularly when the cells make energy, produces a mess known as *reactive oxygen species*. This is similar to kitchen grease or the oily fume that comes off of a hot pan. To cope with those kitchen messes, we turn on the exhaust fan and put on a kitchen apron. Similarly, cells have a timed mechanism to clean up after themselves. This also includes the detoxification process.

- Repair and cell division is circadian. Our body is being repaired and rejuvenated every day. Just like our plumbing gets weaker and leaks after a while, we have hundreds of miles of blood vessels that need to be checked for leakage and repaired. Similarly, our gut lining and skin needs daily repair to keep bacteria, chemicals, and toxins from entering our body. Inside every organ, many cells die and need to be replaced. Our blood cells also need replacement. This repair, through the production of new replacement cells, does not happen randomly; rather it occurs at a specific time of the day: at night, when we're asleep.

- Cell communication is circadian. Our organs need to communicate with each other, and this happens within a distinct rhythm. For example, when we are full, the hormone leptin is produced in the body's fat tissue, sending signals to the brain to stop us from eating more. Similarly, when we eat, hormones from our gut tell the pancreas to produce insulin so that glucose from our food is absorbed into our liver and muscles. These communications are stronger at certain times and they weaken at other times of the day.

- Cell secretion is circadian. Each cell produces something of value for its neighbor or for the whole body. Consequently, every organ produces something that gets into the bloodstream or is delivered to its neighbor. The production and secretion of these molecules are circadian. For example, the liver produces several types of molecules that are necessary for forming blood clots. Since the blood-clotting factors are circadian, if we carefully measure our bleeding time or clotting time, we will see a clear circadian rhythm. This can optimize when we should schedule surgery for quicker healing. Similarly, our nasal lining, gut lining, and lung lining produce lubrication, and this production is also circadian.

- Almost every drug target is circadian. This is one of the most relevant effects of circadian science, especially for people who are undergoing treatment for any chronic disease or cancer. Remember, thousands of genes in an organ turn on or off at a certain time. Imagine if you could target the gene that makes a protein that helps make cholesterol in your liver. That protein has a daily rhythm, making more cholesterol in the morning and less at night. If we want to reduce cholesterol production in the liver, wouldn't it be better to have a drug that will block the cholesterol-making protein when it's most active?

SCN: The Master Clock

Scientists knew that cells communicated with each other, but we wondered if our internal clocks communicated from organ to organ. Scientists found a small cluster of cells that function as a master clock—just like atomic clocks are the master clocks for all other clocks in the world. These cells, collectively known as the *suprachiasmatic nucleus*, or SCN, are strategically located at the hypothalamus, the center of the base of the brain, which houses the command centers for hunger, satiety, sleep, fluid balance, the stress response, and more. The 20,000 cells that make up the SCN are indirectly connected to the pituitary gland, which produces growth hormone; the adrenal glands, which release stress hormones; the thyroid gland, which produces thyroid hormones; and the gonads, which produce reproductive hormones. The SCN is also indirectly connected to the pineal gland, which produces the sleep hormone melatonin.[3]

The SCN's function is so central to daily rhythm that when it is surgically removed, as scientists have done with rodents, the animal loses all of its rhythms. In fact, in the very end stage of neurodegenerative diseases such as Alzheimer's, if the SCN is also degenerated, the patient loses his or her sense of time: They go to bed or stay wide awake, feel very hungry, and go to the bathroom at random times of the day or night.

The SCN is the link between light and timing, because it receives information about light from the outside world and shares it with the rest of the body. The melanopsin cells of the retina make direct connection with the SCN, which is why our master clock is most sensitive to blue light. When the SCN gets reset by light, it resets all the other clocks that are in the hypothalamus: the pituitary gland, adrenal gland, pineal gland, etc. The other clocks in the body, like the liver clock and the gut clock, create their circadian rhythm from a combination of the SCN signal and the timing of the foods we eat. The SCN clock is connected to the hunger center in the brain, so the SCN actually tells the brain when to feel hungry and when not to feel hungry. So, in that way, the SCN guides and instructs us when to eat, which indirectly instructs the liver clock, the gut clock, the heart clock, and so on.

There is a circadian rhythm for drinking water that helps our liver and muscles do many jobs. Liver cells swell up when you eat to make their own protein (liver produces most of our blood protein). The cells can swell only when they take up water. This is why we know that hydration helps organs do the necessary chemical reactions to supply energy and keep their vital functions going.

The system is flexible enough that if food appears at the wrong time, the system resets within a few days. The gut resets itself so that it produces digestive juice just before the food appears, and the liver clock resets to process nutrients that are absorbed in the gut. Slowly, after a week or so, some of the brain clocks are affected. They are reset to the new eating times. In this way you can see how light and when we eat can affect many of these clocks.

The Three Core Rhythms

The clocks in different organs work like an orchestra to create three major rhythms that form the essential foundations of health—sleep, nutrition, and activity. What's more, these rhythms are entirely interrelated and are also under our control. When they all work perfectly, we have ideal health. When one rhythm is thrown off, the others are ultimately upset, creating a downward spiral of poor health.

Your body's rhythms work like a busy intersection that's controlled by traffic lights. Any activity, from the way the brain functions to the way we digest food, works just like the flow of traffic: Each function is coming from one direction, but ultimately everything converges. If we don't have the right traffic pattern, our rhythm is off. Because we can't have all body functions happening at the same time, we either get stuck at an interminable red light or, like cars colliding in a traffic accident, our rhythms will interfere with one another. And when we fail to pay attention to the traffic light, or when we work against the optimal rhythm, it confuses the signals and eventually compromises our health.

Rhythm 1. Sleep: The Myth of Morning Larks and Night Owls

Many people believe that they either go to sleep remarkably early or late and wake up either early or late. They attribute these sleep habits to genetics and then describe themselves as either night owls who can stay up late, or morning larks who wake up early.

In fact, whether you're a night owl or a morning lark can change with age. Infants and young children tend to wake up early because they fall asleep in the first few hours of the night. If you are trying to keep your child awake well past 9:00 or 10:00 p.m., you're actually interfering or disrupting the natural tendency for them to fall asleep. Delaying a child's natural sleep pattern has become an important health problem, and it affects brain development. In fact, even in adults, attention deficit hyperactivity disorder (ADHD) and autism spectrum disorder (ASD) are now linked to going to bed very late in the evening, not getting enough sleep, and then staying indoors most of the day.[4] Of course, children are sometimes kept up later at night because it's only natural for parents to want to spend time with their babies. This is a big problem in India and China, where many parents have long commutes to and from work.

Teens are more likely to go to bed late and wake up late. Many high schoolers can stay awake well past midnight, but they don't get sufficient sleep if they wake up before 7:00 a.m. to go to school.

As we get older and hit our thirties or forties, we naturally revert to being early risers. That means we'll have less difficulty falling asleep in the evening and are likely to wake up at the crack of dawn. However, postpuberty, there is a tendency for females to wake up earlier than males. This difference disappears during middle age, as sex hormones weaken, and it clearly shows how declining sex hormones affect sleeping patterns.[5]

We are programmed to maintain at least a 9-hour sleep pattern when we are babies and a 7-hour sleep pattern for the rest of our lives, yet the overall clock system dampens with age and becomes less effective. As we age, the internal drive to have consolidated sleep or wakefulness slowly breaks down, and we wake up more easily when disturbed by light or sound and have difficulty getting back to sleep. This is when nurturing a body clock with better habits becomes critically important.

While many people think that changes to their sleep cycle are genetic, the chance of having a genetic mutation is remote at best. Very few people have a genetic defect that changes their clock so profoundly that it becomes difficult to adopt new habits to correct it. But studying these people has given us insight into the human circadian rhythm.

One woman named Betty knew she had a sleep problem, and it was so debilitating that she searched for a solution. Betty would get the much-prescribed 7 hours of sleep every day, but the hours she slept were not the norm. Each night she would go to sleep at 7:00 p.m. and wake up at 2:00 a.m. Her sleep routine was a big problem for her, as it limited the amount of time she could spend having a normal social life. Betty went to many sleep doctors, each of whom checked her out and told her that she was fine because she was getting 7 hours of sleep. But no matter how hard she tried, she couldn't adjust her sleep patterns.

The last doctor she saw was Christopher Jones at the University of Utah, who also initially thought Betty's sleep schedule was a nonissue, until Betty told him that several members of her family had the exact same sleep pattern. Chris immediately thought that this might be a genetic mutation within the family. He shared Betty's story with molecular geneticist Louis Ptacek and his wife, molecular biologist Ying-Hui Fu, who saw Betty's problem as a challenge. Over the next few years, Ptacek and Fu found a single change in Betty's Per gene, the same gene that had been altered in Seymour Benzer's and Ron Konopka's mutant fruit fly experiments. For the first time, a single gene mutation in humans was conclusively linked to a change in sleep-wake cycle or circadian rhythm.[6]

This single, highly rare mutation made Betty's circadian clock run faster than normal, and it would always remain in that state. In the morning, when our brain clock gets synchronized with the morning light, the clock starts counting how many hours we are awake. For most people, after 12 hours of being awake our brain clock gently nudges us to begin preparing for sleep.

Most of us will want to go to bed after 16 hours of staying awake. But Betty's clock was running faster. Betty's brain counted 12 hours of wakefulness as 14 hours. And 14 hours after waking up, her brain clock thought she had been awake for 16 hours, and she would find it too difficult to stay awake.

A few years later, Fu discovered another family who had a potentially different mutation in a gene called Dec2, which can reduce sleep need. People with this mutation can sleep for only 5 hours yet wake up feeling completely rested and can carry out their daily routine perfectly fine.[7]

Even if you have a bad gene, healthy habits can often override its deleterious effects. Although Betty had difficulty staying awake late into the night and socializing with friends, other people like her play this genetic twist to their advantage by going to work early so that they can get home earlier, or by lengthening their workday. However, the majority of people, especially late sleepers, may not have a defective gene. Their late sleeping may be caused by other habits that run counter to their circadian code.

I once met a successful businessman who complained how he struggled to fall asleep every night and had difficulty sleeping for a few hours in a row. He was convinced that he must have bad sleep genes. But after talking to him for a few minutes about his daily routines and eating pattern, it became clear to me that his sleep problem was due to the three cups of strong coffee he drank every day between the late afternoon and going to bed. Once he stopped drinking coffee after lunch, he started falling asleep around 10:00 p.m. and was able to stay asleep for 7 full hours.

Another way we know that thinking of yourself as either a morning lark or a night owl is tied to bad habits is shown through an experiment performed by Ken Wright Jr. at the University of Colorado, Boulder. He led a camping trip for a few people who believed they were in the moderate night owl camp—they went to bed late and woke up late every day. Before the trip, they all monitored their sleep patterns and took saliva samples to figure out what time they produced the largest amount of the sleep hormone melatonin. Ken found that many of the night owls had a delayed onset of melatonin production: Their sleep hormone would not rise until 10:00 p.m. and then would peak after midnight.

However, after 2 days of camping in the wild, they again tested when their melatonin was rising. Surprisingly, all the individuals who were completely convinced that they were genetically programmed to be night owls were absolutely normal in terms of their melatonin production, which was now occurring earlier in the evening compared to their initial lab tests before the trip. What's more, they all were able to get to sleep before 10:00 p.m. Their melatonin levels would rise as early as 7:00 or 8:00 in the evening instead of after 9:00 or 10:00 p.m., and they could not stay awake.[8] The change was not due to uncomfortable sleeping arrangements but to the lack of bright light in the evening and the removal of other bad habits, such as working in the evening and late-night caffeine intake. Without access to bright light at night, these people were able to resume a more normal circadian rhythm.

This experiment, among others, is one of the reasons why I'm so convinced that we are masters of our health. Correcting habitual behaviors is the key to improving your circadian code. I've experienced this correction firsthand. While camping in Kenya's Maasai Mara National Reserve with no electrical lighting and surrounded by wild animals, my colleagues and I had no incentive to stay awake late into the night. I had some of the best sleep in years and woke up refreshed at least 30 minutes before sunrise for several days in a row. When I returned to San Diego, my old sleeping pattern returned—going to bed late at night and trying to wake up an hour after sunrise. When I shared this story with my colleagues, they pointed out many differences in my San Diego lifestyle versus how I lived in Maasai Mara: In Kenya, I was exposed to lots of light during the day, no light at night, less noise, relatively cooler temperatures at night, and earlier dinner. Each of these factors has been shown to contribute to better sleep.

Rhythm 2. When You Eat Affects Your Clock

If the circadian system's main goal is to optimize energy intake and survival, then what happens to the system when food is available at the wrong time? For rodents, if food is available only during the daytime (when they are supposed to be asleep and fasting), what happens? Does the SCN master clock ignore the food? This would have a detrimental effect on their health to say the least because if they choose to ignore the food cue they will die. In fact, when mice learn that food is available only during the day, they start waking up an hour before food arrives to search for food. In other words, they figure out a mechanism to anticipate food. But after eating their food, they go back to sleep (as they usually do during the day), and they stroll around at night. In other words, their SCN clock, controlling the daily sleep-wake cycle, continues to work fine, except for a brief period during the day when they wake up to eat food.

When mice eat food during the daytime when they are not supposed to eat, what happens to the food? Does it get digested and metabolized in the liver, where the liver clock regulates metabolism? This was a conundrum. Up to that time, we believed that while the liver may have a clock, its function was at least partly controlled by the brain, which sends a signal to the liver. Yet we were kind of skeptical at the same time, because it would take a lot of energy and effort for the liver clock to be so reliant on the brain. Besides, if the animal eats every day at the wrong time (daytime for our mice), yet the liver clock is programmed to metabolize food at night, then the liver clock would not be able to metabolize the food eaten during the daytime.

So, in 2009, we did a simple experiment. We took some mice, which are usually nocturnal, and we fed them only during the daytime. Then we looked at their liver function. We saw that almost every liver gene that turns on and off within 24 hours completely tracked the food and ignored the timing of light exposure.[9] That meant that it was the food that reset the liver clock, not the brain.

That finding completely changed how we think about circadian rhythm in relation to light and food. Instead of thinking that all timing information from the outside world to every organ in our body has to go through the blue light sensors, we now knew that the body was able to sync to other cues. Just like the first light of the morning resets our brain clock, the first bite of the day resets our organ clocks. In fact, food timing can be a powerful cue to override the master signal from the SCN master clock.

Think about your morning breakfast. Have you ever noticed you feel hungry around the same time every morning regardless of what you ate for dinner the night before? This happens because our brain clock or the clock in the hunger center tells us when we should be hungry. At the same time, the brain and gut talk to each other, and the clock in the gut tells the brain to get ready for a rush of breakfast. The pancreas is also ready to secrete some insulin, the muscles are ready to soak up some sugar, and the liver is ready to store some glycogen and make some fat and send it off for storage.

If you typically eat breakfast at 8:00 a.m., you've set an appointment with your stomach, liver, muscles, pancreas, etc., and they will be ready to process the breakfast at 8:00. This first bite is also one of your clock's links to the outside world: Breakfast becomes the cue that syncs the internal clock with the outside time. As long as you eat breakfast at 8:00 (give or take a few minutes), your internal clock will be in sync with the outside world.

But imagine that one day you have to get up early to go to catch a flight from Los Angeles to Chicago and your schedule is disrupted. Instead of eating at 8:00 a.m., you "need" to eat at 6:00: After all, you've been taught that breakfast is the "most important meal of the day." When you sit down in front of your cereal bowl, you'll likely notice that you're not really feeling hungry. That's because your brain hasn't sent the signal to your stomach to get the digestive juices ready to process your food. Your liver and the rest of your organs aren't ready, either.

But no matter: You know best, so you eat anyway. With the first bite, your stomach will switch on its emergency mode and process the food. The body has to drop everything it is supposed to do at 6:00 a.m. and turn attention to the incoming food. Or it can ignore your food and it will stay undigested for a couple of hours. Typically, the body chooses the first option: It stops its usual before-breakfast activities, which include cleansing itself and running on stored energy. So, when the early breakfast appears, the body has to drop cleaning up and turn off the fat-burning switch so it can use the fresh food that you just ate as fuel.

What's more, the clocks in your stomach, liver, muscles, pancreas, etc., will also take note of the unanticipated breakfast and will get confused. These clocks will think perhaps they are wrong and that it is 8:00 a.m. To make up for the "lost time," the clocks in these organs will try to speed up. But your circadian clock has many moving parts, and it is not that easy to speed all the clocks in different organs so quickly and get them back into alignment. Usually, they can adjust themselves by an hour per day.

When you show up for breakfast the next day, it's 8:00 a.m. in Chicago but your body still thinks it's 6:00 a.m. in Los Angeles, and the stomach is still not ready. It goes to emergency mode and tries to process your food. And again it tries to speed up the clocks.

By day 4, you've created an entirely new circadian code that has adjusted to your schedule. But guess what? It's time to go home. When you get back to Los Angeles and sit down for breakfast at 8:00 a.m., your system thinks it's 10:00. This time, the organs were ready for your breakfast appointment at 6:00, but they didn't get any food. So they started the next task on their lists. As soon as you eat your breakfast, your stomach, liver, muscles, pancreas, etc., have to now drop what they were doing or turn their attention to processing your breakfast. This time, they choose to multitask. Again, the clocks try to reset to the new breakfast time by slowing down for the next few days.

This example shows how an erratic breakfast time confuses your organs and compromises their function. Using the circadian clock, each organ is programmed to process food for a few hours starting from breakfast. If your breakfast was at 8:00 a.m., then the system works optimally for about 8 to 10 hours. Every time we eat, the entire process of digestion, absorption, and metabolism takes a couple of hours to complete. Even a small bite of food takes an hour or two to be processed. After about a 10-hour window, the gut and metabolic organs will continue to work on your food, but their efficiency slowly goes down as they are not programmed to be open for business 24-7. The clocks in different organs are not that efficient. Your gastric juice and gut hormones are produced at a different rate and your digestion slows down, giving you a sense of indigestion or acid reflux.

What's more, just like a late breakfast interferes with the other tasks your organs have to accomplish, a late dinner will, too. This time, the disruption is more severe. The same food that would have taken a couple of hours to digest at 6:00 p.m. takes longer to digest at 8:00 because you are outside of that optimal 10-hour window. This extra work interferes with the next task by delaying or even completely removing that task from the list.

Now you might be thinking, *Dr. Panda, who cares? I'm sleeping anyway.* But here's the rub: Our cells cannot make and break up body fat at the same time. Every time we eat, the fat-making program turns on and the cells in our liver and muscles create some fat and store it. The fat-burning program slowly turns on only after the organs realize no more food is coming their way, and that takes a few hours after your last meal. And it takes a few more hours to deplete a good portion of stored body fat.

Suppose you had your dinner at 8:00 p.m. and you finished a half hour later. The clock is continuing to tick, your fat-making process is slowly winding down, and at around 10:30, you have an urge to snack. A piece of fruit, a bowl of cereal, a granola bar, a handful of nuts, it doesn't matter. As soon as that food gets into your stomach, the stomach clock that had already put the "kitchen is closed" sign up has to get back to work and process your snack. That same food in the morning would have been processed in an hour or so, but now the stomach wasn't prepared for food, and it will take a few hours to process the snack. Your fat-making process continues past midnight and the fat-burning process won't begin until the morning, but when you eat your breakfast, the switch turns again toward making fat.

Sitting here in my lab, I can imagine you scratching your head again: *Dr. Panda, what's the big deal? Aren't we talking about just a few ounces of fat gain after a late-night snack? Won't my metabolic rhythm come back the* *next day?* Actually, it's worse than you think. It is hard enough for the body to monitor hormones, genes, and clocks for someone with a strict eating routine. But when eating occurs at random times throughout the day and night, the fat-making process stays on all the time. At the same time, glucose created from digested carbohydrates floods our blood and the liver becomes inefficient in its ability to absorb glucose. If this continues for a few days, blood glucose continues to rise and reaches the danger zone of prediabetes or diabetes.

So, if you've wondered why diets haven't worked for you before, timing might be the reason. Even if you were diligently exercising; counting calories; avoiding fats, carbs, and sweets; and piling on the protein, it's quite likely that you weren't respecting your circadian clocks. If you eat late at night or start breakfast at a wildly different time each morning, you are constantly throwing your body out of sync. Don't worry, the fix is equally simple: Just set an eating routine and stick to it. Timing is everything.

Rhythm 3. The Effect of Physical Activity on Timing

When we're not eating or sleeping, we're supposed to be engaged is some form of physical activity. In fact, our metabolism and physiology evolved so that our body can perform physical activity throughout our wakeful hours from morning to evening. When we are active, we are supposed to use most of our muscles, which together constitute nearly 50 percent of our body weight. Many of our muscle groups are under autonomic control and work without us even knowing it. These include the cardiac muscles of the heart and the smooth muscles of our digestive tract. Yet even these muscles have a circadian rhythm; they are more efficient during the day than at night.

Our gut muscles automatically stretch and flex to produce what we call *gut motility.* This is what moves digested food from the stomach through the intestines. Gut motility increases during the daytime and is very slow at night. Because gut motility is less active at night, when we eat late, the food moves slowly down the tract and we can develop indigestion.

Our lungs and heart are both muscles that have a circadian variation— we have a relatively higher heart rate and heavier breathing during the day, and both slow down at night. The higher heart rate and breathing help distribute oxygen and nutrients throughout our body, including to our muscles during the day, priming us for physical activity. At night our muscles don't need the same levels of nutrients and oxygen as they do during the day, when we are more likely to use them. This may be one reason why heart rate and breathing slow down at night, which helps the body cool down so we can sleep better.

Most of the muscles are activated when we do physical activity. Physical activity has immense benefits for health, and some activity may have an effect on the circadian clock. Some of the earliest experiments examining the effect of physical activity on circadian rhythm were done on mice that had free access to an exercise wheel. When these mice were allowed to hop on the exercise wheel whenever they wanted, they voluntarily ran on the wheel every night. Researchers found that exercising mice have a robust circadian clock; they sleep better when they are supposed to and are less sleepy when they are meant to stay awake.[10] The effect of physical activity on sleep did not seem to involve food and didn't affect their thirst.

This early observation has prompted several human studies involving a range of participants, from teenagers to older adults. All of the studies have come to the same conclusion: Physical activity improves sleep. Among teenagers, vigorous physical activity not only improved how quickly they fell asleep or how well they slept, it also improved their mood during the day, increased concentration, and reduced levels of anxiety and depressive symptoms.[11] Among older adults (50 to 75 years old), moderate physical activity or even regular stretching improved sleep onset, sleep quality, and sleep duration and reduced dependence on sleep medications. Older adults with moderate physical activity also had fewer episodes of feeling sleepy during regular daytime activity.[12,13,14] When the timing of our sleep improves, our circadian rhythm improves.

What Counts as Physical Activity?

Any form of movement that results in energy expenditure is thought of as the physical activity of day-to-day living. Physical fitness is the ability to perform physical activity. Participating in sports is one form of physical activity that is competitive and involves thinking and planning. General exercise is another form of physical activity that is planned and structured, and it is defined by its frequency, duration, and intensity. Gardening, moving heavy items, leisurely walking, and doing household chores also count as physical activity. Chapter 7 includes a table that lists all kinds of physical activities and how they rate as compared to each other.

CHAPTER 3

Track and Test:
Is Your Circadian Code in Sync?

Life expectancy for a baby born in 1900 was only 47 years.[1] Only 1 in 100 would live beyond 90 years old, and a third would die before the age of 5. Infectious diseases, caused by germs and other bacteria, were the major health challenges. Scientists fought those diseases with better sanitation, vaccines, and antibiotics that saved lives. Now, a typical newborn in the Western world is expected to live up to 80 years. Nearly all of us will suffer from one or more chronic diseases, including diabetes, obesity, heart disease, depression, or anxiety. The causes for these diseases are not likely to be infections; instead, they relate directly to the bad lifestyle choices we make. And the medications we take only manage our symptoms; there are no assured medical cures for most of these diseases. The medications work best when combined with making better, healthier lifestyle choices. And those choices are linked to your circadian code.

Experts usually define a healthy lifestyle by the type of food we are supposed to eat and the type of exercise we should do. I want you to focus not so much on the *what* of a healthy lifestyle but on the *when*. A healthy lifestyle includes what and when you eat, when and how much you sleep, and when and how often you move. By focusing on the when, you are harnessing the power of your circadian code, which can compensate for those times when you make less than exemplary choices. Better

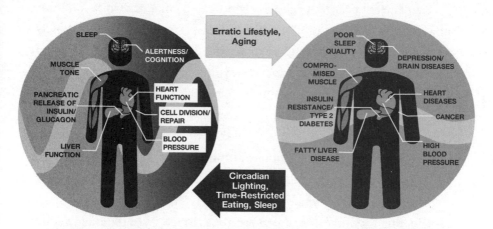

Erratic lifestyle or aging promotes circadian disruption and various diseases. Circadian lighting, time-restricted eating, and restorative sleep sustain our circadian rhythm and prevent or reverse these diseases.

still, by living in alignment with this internal rhythm, you reap even greater benefits that come along with making good lifestyle choices.

How Strong Is Your Code?

More likely than not, we are born with a strong circadian clock that instructs every aspect of our body to work efficiently. It sets a daily rhythm for when to sleep and wake up, eat, and be active. We are at our best health when we are living at a pace that is aligned with this perfect rhythm. However, sometimes life gets in the way. As you've learned, it is less likely that your genes are sending your clock the wrong message and disrupting your rhythm and more likely that your habits have screwed with your body clock. And unfortunately, it does not take too many days of disruption to completely throw off our rhythm. When we consistently shift our work or stay awake late into the night or eat outside our essential routine, it confuses our circadian rhythms, which will ultimately lead to poor mental and physical health.

Our current health can also affect how well our clock works, and this influence can be direct or indirect. For example, depression often affects our sleep-wake cycle, leading to either increased insomnia or increased sleep. It also makes people want to stay indoors in gloomy, dark rooms. Both of these symptoms disrupt our clock by throwing off both light and timing, which then pushes us further down the spiral of depression. Chronic diseases such as type 2 diabetes or liver disease occur when large numbers of genes that should turn on and off at different times of the day get stuck in either the on or off position, leading to blood sugar dysregulation, binge eating, or food cravings. Breaking this cycle with a better eating rhythm can nudge these genes to return to their daily cycle and reverse these diseases. Finally, the body produces many chemical signals as it fights against the tumor in certain cancers. Some of these signals can travel in our bloodstream to distant organs, where they disrupt normal rhythmic functions. Again, once we can adapt to a lifestyle that is aligned with the natural circadian rhythm by maintaining the proper sleep-wake or eating-fasting rhythms, we can counter the disrupting signals and help accelerate recovery.[2]

We're Just Not That Resilient

You may think that having a bad night's sleep, working until all hours of the night, or eating a big meal in the middle of the night won't kill you. Well, to some degree, you're right: A one-off experience is not likely to do much damage. However, bad lifestyle habits directly affect our circadian code, and while they may not kill us in the strictest sense, they do make us vulnerable to factors that can kill us. For instance, in a study simulating jet lag or shift work, when mice were put on a shift-work-like schedule by simply changing the time when lights turn on and off by a few hours, in just a few weeks the mice became so frail and their immune system so weak that they succumbed to infection, and if left untreated half of them died.[3] Similar results have been found in human studies. In a large study of more than 8,000 workers from 40 different organizations, researchers found that shift workers were more likely to suffer from infectious diseases, ranging from the common cold to stomach infections, than non–shift workers.[4] These observations show us that when our rhythm is off and we come in contact with everyday bugs or viruses that we are typically resistant to, they can cause serious illness.

It takes longer than you may imagine for your body to adjust to even the smallest glitch in your circadian rhythm. For example, one night of shift work can throw off your cognitive abilities for an entire week. Cross-country travel may seem benign, but when you have to adjust to a new time zone, you may feel jet-lagged for a few days. For most people, our circadian clock takes almost 1 day to adjust to each hour of time-zone shifting; for some people, it can take 2 days per hour of time shift. Similarly, when you stay awake for 3 extra hours and delay your breakfast by 3 hours on the weekend, it affects your body in the same way as flying from Los Angeles to New York. That's why partying or staying awake late into the night is the same as flying over a few time zones; hence clock scientists call this habit *social jet lag*.

You can figure out how well you adjust to jet lag by noticing how many days it takes your body to get used to daylight saving time, when our clock shifts by just an hour. Now you can see what happens when we socialize for just a few nights every month, staying up past our normal bedtime.

And it's not just a change in sleep that can throw off your clock: A change to any of the three core rhythms—sleep, timing of food, and activity—can affect any one of your organs in a similar way. Just as the circadian code regulates a range of functions in different organ systems, interfering with this primordial rhythm can compromise optimum function of any organ. Unlike smoking, which we know has a direct effect on your risk for lung cancer, circadian rhythm disruption does not lead to one specific disease, but it can compromise health in many different ways. If you are already susceptible to one specific type of illness, you may notice its symptoms first. It's like if you took five different car models off-roading for the day; each car will come back with its own unique set of problems based on its model. Some will come back with their tires intact but their suspension will be off. Some will have transmission problems or an alignment issue. So, if you have always had problems with acne, disruption in your circadian rhythm might lead to a major breakout. If you have a sensitive stomach, circadian disruption can trigger heartburn or indigestion.

It's quite possible that some of the daily discomforts, frequent illnesses, or chronic diseases you may have are linked to circadian disruption. The symptoms of many illnesses include poor or excessive sleep,

Circadian Disruption Affects Health Over Time

Short-term circadian rhythm disruption
(1–7 days)

Sleepiness/insomnia, lack of focus, migraine,
irritation, fatigue, moodiness, indigestion, constipation,
muscle ache, stomach pain, bloated stomach,
blood glucose rise, susceptibility to infection

Chronic circadian rhythm disruption
(weeks, months, or years)
in combination with genetic
predisposition/poor nutrition

Gut diseases, immune diseases,
metabolic diseases, affective or mood diseases,
neurodegenerative diseases, reproductive diseases,
chronic inflammation, various cancers

The longer your circadian rhythm is out of sync,
the greater the risk of developing a serious disease.

change in appetite, or reduced physical activity. These are all disruptions to your circadian code. However, by fixing your rhythm you can potentially correct your disease or lessen its severity. This is why I say that nurturing your circadian rhythm acts as a grand corrector of all maladies.

If you are experiencing any physical discomfort, or changes to the way you think, don't ignore them: They are early warning signs of chronic illness. First, pay attention to whether your daily sleep, activity, and eating patterns have changed. Try to return to your normal patterns. If symptoms persist, see a doctor.

Discomforts can slowly take shape as diseases that require prescription medications. Medications for chronic disease don't work like medications for bacterial infections—a course of antibiotics will kill a pathogenic bacteria and you are cured. Chronic diseases cannot be cured and are largely managed by taking medications for the rest of your life. At the same time, you have to cope with the adverse effects of these medications. A recent review of the top-ten highest grossing drugs in the United States showed that for every person they help, they fail to treat between 3 and 24 people.[5] Worse, circadian disruption over time makes treating symptoms less effective. It can slow down recovery or even make you resistant to therapy. For example, women who undergo treatment for breast cancer and cannot fall asleep at their preferred bedtime have lower survivorship rates than those who maintain a consistent bedtime.[6]

In Part III, you'll learn about how living in opposition to each of the core rhythms affects your microbiome, your metabolism, your immune system, and your brain, and what you can do to reverse these areas of poor health. These are what I like to call the "big-ticket" items of health. However, messing with your circadian code for even a few days makes life rougher. It will come as no surprise that sleep-deprived people are more unpleasant and even hostile in their social interactions than those who get adequate sleep. For instance, a 2011 study showed that positive emotions decrease and negative emotions increase in response to sleep deprivation, particularly in adolescents.[7] Poor sleep compromises our ability to evaluate negative or positive rewards and make rational decisions. It also affects our ability to attend to the task at hand. This affects both our work life and our home life. One 2017 study showed that when sleep-deprived couples argue, they are more irrational or they perceive the other person as more irrational.[8]

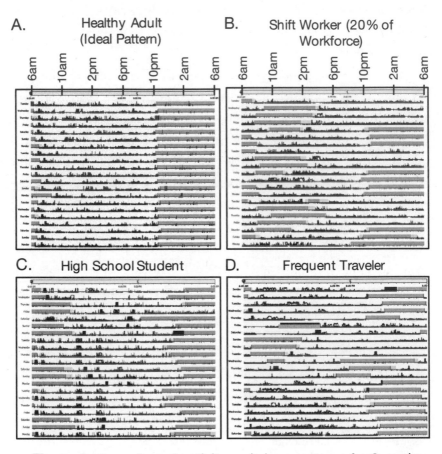

These charts represent activity and sleep patterns for 3 weeks of (a) a healthy adult with ideal activity-sleep pattern, (b) a shift worker, (c) a high school student, and (d) a frequent traveler. Each horizontal line reports activity (in dark spikes) and sleep (gray bars). Notice the healthy adult goes to bed around 10:30 p.m. every night and gets 8 hours of sleep. All the other people have extremely delayed sleep time at least one day every week.

And as I've said, when one rhythm goes, so go the rest. All of us have experienced this during real or social jet lag, like when we attend a late-night party. Let's say you go out dancing. The light you're exposed to suppresses your need for sleep. Every hour you stay awake past midnight disrupts your circadian rhythm more. The next morning, you feel terrible. You're tired, and it's hard to catch up on all of the lost sleep the next day even if you can sleep in. Waking up late also disrupts your regular eating pattern, moving it from 8:00 to 10:00 a.m., and it may also eat into your exercise time. You might also find that your brain is foggy and it's harder to pay attention or make even simple decisions.

Sometimes, chronic diseases cause their own circadian disruptions. For example, obesity increases the risk for obstructive sleep apnea. Not having restorative sleep because you can't breathe freely increases sleepiness during the day and reduces the drive to exercise. As physical activity goes down, the drive to sleep at night also reduces. This keeps the person awake late into the night under bright light, and as they stay awake, there's a greater likelihood that they'll continue to eat late into the night.

Testing Your Circadian Code

If you are living with a medical condition, it is important to know if your condition may be disrupting of your daily rhythms. I have developed two quizzes that you can take in the privacy of your own home to help you see if the quality of your circadian code is affecting your health. The first quiz focuses on the way you think and feel right now. The second offers an opportunity for you to track how far off you are from living within an optimal rhythm.

The Circadian Code Health Assessment

The first quiz presents various symptoms and conditions you may be suffering from. These symptoms or conditions might affect your sleep or eating cycles. Or they may be a response to a poorly operating circadian code. Either way, recognizing that you have these issues, and that they may be affecting you more than you realize, is the first step to addressing them.

If your work requires you to stay awake for at least 3 hours between 10:00 p.m. and 5:00 a.m. for more than 50 days a year (once a week), you are a shift worker and at risk of suffering from shift-work-related diseases. Many of us do not sleep enough, eat randomly, do not participate in significant physical activity, or do high-intensity physical activity at the wrong time. Understanding which factors affect your clock will help you make small changes that will add years of good health to your life.

Answer the following questions truthfully, circling the correct response. The results will be as individual as you are: There are no right or wrong answers. However, if you answer "yes" to any of these questions, it is likely that optimizing your circadian system will benefit your health. Don't worry if you aren't perfect; almost everyone has room for improvement.

PHYSICAL HEALTH

Has your doctor told you that you are overweight?	Y/N
Have you been diagnosed with either prediabetes or diabetes?	Y/N
Are you taking *prescription* medication for a chronic disease, such as heart disease, blood pressure, cholesterol, asthma, acid reflux, joint pain, or insomnia?	Y/N
Are you taking *over-the-counter* remedies for acid reflux, pain, allergies, or insomnia?	Y/N
Do you have an irregular menstrual cycle?	Y/N
Do you have hot flashes or disrupted sleep related to menopause?	Y/N
Do you have a decreased libido?	Y/N

Have you been diagnosed with a disease linked to chronic inflammation, such as multiple sclerosis or inflammatory bowel disease?   Y/N

Do you have frequent lower back pain?   Y/N

Have you been diagnosed with sleep apnea?   Y/N

Do you snore?   Y/N

Do you wake up feeling congested or with a stuffy nose?   Y/N

Do you have frequent abdominal pain, heartburn, or indigestion?   Y/N

Do you have frequent headaches or migraines?   Y/N

Do your eyes feel tired at the end of the day?   Y/N

MENTAL HEALTH

Do you feel anxious?   Y/N

Do you feel low or have frequent blue moods?   Y/N

Do you struggle with attention and focus?   Y/N

Do you experience brain fog or poor concentration?   Y/N

Do you frequently lose items, like your glasses, a charging cable, or keys?   Y/N

Are you forgetful of names and faces?   Y/N

Do you rely on a calendar or to-do lists?   Y/N

Do you get tired in the afternoon?   Y/N

Do you wake feeling tired?   Y/N

Have you been diagnosed with post-traumatic
stress disorder (PTSD)?   Y/N

Have you been diagnosed with attention deficit
hyperactivity disorder (ADHD), autism spectrum
disorder (ASD), or bipolar disorder?   Y/N

Do you have food cravings?   Y/N

Do you feel like you have a lack of willpower over food?   Y/N

Have you been told that you are irritable?   Y/N

Do you have trouble making decisions?   Y/N

BEHAVIORAL HABITS

Do you take less than 5,000 steps a day?   Y/N

Do you spend less than an hour outdoors under
daylight each day?   Y/N

Do you exercise after 9:00 p.m.?   Y/N

Do you spend more than an hour on the computer,
your phone, or watching TV before bedtime?   Y/N

Do you have one or more alcoholic drinks (cocktails,
wine, or beer) after dinner?   Y/N

Do you forget to drink water throughout the day?   Y/N

Do you drink coffee, tea, or caffeinated soda in the
afternoon or evening?   Y/N

Do you consume chocolates, high-carb foods (doughnuts,
pizza), or energy drinks to improve your energy level?   Y/N

Do you binge on foods late in the day regardless of hunger? Y/N

Do you drink or eat anything (other than water)
after 7:00 p.m.?   Y/N

Do you sleep with a light on?   Y/N

Do you set aside less than 7 hours for sleep and rest every day?   Y/N

Do you need an alarm clock to wake up in the morning?   Y/N

Do you typically catch up on sleep on the weekends?   Y/N

Do you eat whenever food is presented to you, even
if you are not hungry?   Y/N

Assessing Your Responses

Most of us will have a few "yes" answers to the above questions. It is common (but not normal) as we are all shift workers and we do have some circadian disruption in our lives. In the physical and mental health sections, many of us may answer yes to one or two questions, but answering three or more in each section is a sign that your circadian rhythms may not be optimal. You may also assume that some of the symptoms may be benign and negligible because many people of your age or your peers may have the same symptoms. But what is common is not always normal.

In the behavioral habits section, any yes answer is a potential disruption to your circadian clock. Many people typically have five or more yes answers, which means they have many different ways to optimize their rhythm and stay healthy.

Identifying Your Circadian Numbers

The second part of the test is less a quiz and more a tracking exercise. For the next week, fill out the following chart, using the descriptions as your guide. Answering these six questions for a week will give you a fair idea of your daily rhythm. Most likely, you will find that your answers depend on many factors: whether or not you are at work, if it is a weekday or a weekend, or if your lifestyle is too unpredictable. In my lab, we have monitored thousands of people all over the world, and the trend is the same—most people have different rhythms on weekdays and weekends. However, we know that this doesn't have to be the case: When we study the data from indigenous peoples, like the Toba in Argentina or Hadza in Tanzania, their sleep and physical activity patterns are extremely predictable and consistent from day to day.

I can't tell you what your ideal circadian code should be, but chances are you already know it. More likely than not, it is revealed when you are taking a week-long vacation. If you don't drink too much and continue exercising, you may revert back to the rhythm that your body generally prefers. You may find yourself going to bed earlier, being more active during the day, and having fewer cravings for late-night snacks.

However, once you're back from vacation, the realities of your life prevent you from maintaining an optimal circadian code. You may not be able to immediately incorporate every aspect of your ideal rhythm in everyday life, but paying attention to your natural rhythms will give you a good understanding of what you may be doing wrong and help you see the places where you can make small fixes that yield big results. You may also see that you have a choice when it comes to the way you live. Once you know your code, you may choose to stick with the current lifestyle that *you think* is unavoidable for job and/or family, which may lead to slowly succumbing to chronic diseases. Or you can make the decision that your health is more important to you and your family and make some adjustments so that you can be productive at work and home for years to come.

In the following chapters, we'll discuss goals to work toward. But for now, let's take a brief look at the importance of each of these questions.

When (and How) You Wake Up Is the Most Important Event of the Day
When you wake up, open your eyes, and get out of your bed, that first ray of bright light entering your eyes activates the melanopsin light sensors

in your retina that tell your SCN master clock that morning has arrived. Just like in a spy movie when two agents begin their mission and synchronize their watches, seeing the first ray of bright light signals the SCN to set its own clock time to the morning. Usually, when the SCN clock registers morning, it will automatically nudge you to wake up— like an internal alarm clock. But if you need a real alarm clock to wake yourself up, your SCN isn't ready and still thinks it is night. This is why the goal is to be less reliant on an alarm clock and to get enough hours of sleep so that you wake up when your SCN recognizes morning.

As you fill in the chart on page 60, note not only what time you woke up but if you required an alarm clock. We don't wake up the same way our ancestors did even a century ago. In the past, when our circadian clock was in sync with the day-night cycle and we went to bed before 10:00 p.m., our SCN clock would wake us up around dawn. That's when you naturally stop producing melatonin and your sleep drive reduces. Dawn also brings many environmental signals to wake us up, like the first light and the noises of birds and animals. If those cues didn't do the trick, a rise in body temperature would wake you up; as melatonin levels drop to reduce the drive to sleep, cortisol levels go up, making you feel noticeably warmer.

Today, we rarely wake up to these cues. Sleeping in a perfectly temperature-controlled bedroom with double-pane windows covered with thick curtains or blackout shades, we've all but cut out the natural morning signals of sound, light, and temperature. And when we go to sleep late, our sleep drive and melatonin levels are still high at dawn. This is why so many of us require a bone-jarring alarm clock.

Your First Bite/Sip of the Day

Just like the first sight of light syncs your brain clock with light, the first bite of food signals the start of the day for the rest of the clocks in your body. In our research, we found that 80 percent of people eat or drink something other than water within an hour of waking up.[9] The next 10 percent have something within 2 hours, and only a small fraction wait for more than 2 hours to eat. Many people also reported that they often skip breakfast. These numbers just didn't add up, so we dug deeper and found out that the term *breakfast* is clearly misunderstood.

Breakfast means "breaking the fast": the time that passed the night before when you weren't eating or drinking. But what constitutes a true break in a fast? The answer is whatever triggers the stomach, liver, muscles, brain, and rest of the body to think the fast has been broken. And that answer is anything you eat or drink besides water.

You may think that a small cup of coffee with a little cream and sugar is not going to break a fast; most people simply associate their morning brew with an attempt to wake up the brain. Actually, as soon as we put calories in our mouth, our stomach begins to secrete gastric juice in anticipation of digesting food. Then a cascade of hormones, enzymes, and genes start their regular chores. That first cup of coffee or tea is all it takes to reset the stomach and brain clock.

Most of our respondents consumed less than one-quarter of their total daily intake of calories between 4:00 a.m. and noon, while they ate more than 30 percent of their daily intake at night.[10,11] They reported that they were skipping breakfast, but in reality they were just skipping a big meal in the morning. Instead, they were eating a small snack or coffee/tea/juice/yogurt/etc., which they didn't consider a meal. However, our stomach does consider it a meal. It does not matter whether it is a cup of coffee or a whole bowl of cereal, when you break your fast, write down the time.

The End of Your Last Meal/Drink

Just like your brain must switch from being active to resting and rejuvenating at night, your metabolic organs also need to wind down and rest for many hours. The last bite or last sip of the day signals our body to prepare to wind down, cleanse, and rejuvenate. It takes a few hours for the brain and body to get the message and start the process; it needs to be completely sure that no more calories are coming its way. So, just like a cup of coffee starts your metabolic clock, your last bite of food or drink has to be part of the digestive process for 2 to 3 hours before the body can begin its repair and rejuvenation mode.

Culture is one of the biggest predictors of eating patterns. Although many people in the United States eat dinner early, we live in a culture of postdinner, late-night snacking. In many Eastern countries and in parts of Europe, late-night eating is the norm. In some countries, restaurants don't even open for dinner before 9:00 p.m. In some places, late-night dinner is the biggest meal of the day, while in others it is usually a small meal or leftovers from lunch.

Be honest and write down when you took your last bite or last sip (other than water and medication) of the day. You may go into this exercise thinking that you already have a schedule, but my research shows that it is likely more often than not that you don't. We use food to keep us energized or to unwind. The weekends pose a different challenge, as we are often on a completely different schedule, socializing well into the night. Keeping track will show you clearly whether you're adhering to a pattern.

When Do You Go to Sleep?

This is again a difficult question to answer. Wake-up time is relatively fixed by our work schedule, so your bedtime often determines how many hours of sleep you'll get. Some of us have a fixed schedule during workdays. Some may have a fixed bedtime every day but wake up at different times on workdays and off days. The most accurate answer is the time when you have shut off lights, checked your last e-mail/text/social media account, and are in bed with your eyes closed.

What Time Do You Shut Off All Screens?

Just 50 years ago, when someone left the living room, he or she was unplugging from social interaction and entertainment to relax and sleep.

Even in the early years of television, there was not much late-night programming; many TV stations used to go off the air after their prime-time shows. But with the 24-7 social media, television, and streaming entertainment on digital devices, it becomes important to know when you are off the virtual party.

Once we shut off our devices, our brain takes many minutes to unwind. Our eyes receive a big share of light from digital screens, so turning off the screen is also signaling when we turn off all light input to our brain.

What Time Do You Exercise?

There are distinct effects of time of exercise or intense physical activity on circadian rhythm and sleep. So, what time you exercise matters.

	What time did you wake up? With or without an alarm clock?	What time did you go to sleep?	What time did you take your first bite/sip of the day?	What time did you take your last bite/sip of the day?	What time did you shut off all screens?	What time did you exercise?
Monday	Time: Alarm?					
Tuesday	Time: Alarm?					
Wednesday	Time: Alarm?					
Thursday	Time: Alarm?					
Friday	Time: Alarm?					
Saturday	Time: Alarm?					
Sunday	Time: Alarm?					

Assess Your Responses

These six times will give you a good idea about your existing circadian system. There is no magic schedule that will work for everyone. However, use the following information to determine where to start making changes. Even small changes will go a long way in making you healthy, productive, and free of diseases. The first four categories are more important and are relevant to almost everyone, whether you routinely use digital screens or exercise.

- If all six times change by +/– 2 hours or more over the course of the week (between workdays and off days), you have a lot of room for improvement. You will easily find at least one category to fix. Sometimes fixing one category will automatically bring a few others to within an appropriate range.

- Look at the total number of hours you spend sleeping each day: The National Sleep Foundation recommends adults should get at least 7 hours a night, and children require at least 9 hours.[12,13,14] If you are sleeping less and you wake up feeling tired in the morning, the first thing you should work on is getting to bed earlier or figuring out a schedule that allows for at least 30 more minutes of sleep in the morning. If you sleep more than 7 hours and still feel sleepy when you wake, perhaps your sleep quality is not up to mark. Remember, just 3 out of 7 days of poor sleep will throw off your best efforts.

- Look at the total number of hours your stomach is at work: Take the earliest you eat any day of the week and the latest you eat any day of the week, ignoring only one outlier number outside your "normal routine." That is the time period your gut is most likely staying ready to process food. If this number is more than 12, here's the good news: You have something to work on and it will have one of the biggest impacts on your health for the rest of your life. And you're not alone: Only 10 percent of adults eat for 12 hours or less on a consistent basis without following a program such as this. People who can eat all of their food within an 8- to 11-hour window most days will reap the most health benefits.

- Compare your last bite/sip time with your bedtime. The difference should ideally be 3 hours or more.

Is that all? Yes it is. You may be surprised or even a little miffed that adjusting these few things can fix your health. What about counting calories? What about a low-carb, sugar-free, Paleo, vegan, Mediterranean, Blue Zones, Atkins, or Warrior diet? What about critical supplements like fish oil or green tea? You no longer have to worry about them. Just pause for a second and think—only a hundred years ago people all over the world ate different types of foods depending on where they lived: There was no Chinese takeout in New York or bagels in India. And there were no correlated chronic diseases to any particular type of cuisine, whether it was high in fat, carbs, or protein. But our ancestors all over the world had one thing in common—they ate less, did more physical activity, slept more, and completed their daily routines with clockwork precision because they did not have the luxury of light. Again, timing is everything.

In fact, timing is the grand corrector for other behavioral habits. We have clinically seen that when people try to eat all their calories within 8, 10, or 12 hours, they also tap into the wisdom of our circadian body and brain. A natural control over calorie intake sets in, and as much as they try, they cannot actually stuff themselves with too much food in a shorter period of time. This means that as you get used to a smaller window of eating, you'll feel more satisfied with less food.

As we like to say, good habits beget more good habits. After a few weeks, people begin to make better food choices. The usual cookies don't look as appealing, and deep-fried foods don't seem appetizing. What's more, as hormonal balance is restored, your immune system, mood, sleep, happiness, and libido might improve. If you are taking medications to treat your blood pressure, cholesterol, or blood sugar, fixing your circadian code will also improve recovery, and you may need a lesser dose to stay healthy.

Join Our Team

I have created an app that provides information and allows you to easily track your circadian code. Go to mycircadianclock.org to sign up to participate in a 14-week research study and get the free myCircadianClock app for your phone. It's a great way to see a more detailed level of your eating and sleeping habits. As the field of circadian rhythm progresses and we make discoveries in clinical science and public health, we will routinely transmit the new information to our users through the app and through our blog posts.

To participate, all you need to do is record everything you eat and drink, including water and medicines, by simply snapping a photograph and uploading it via the app. You may also record your sleep, or you can pair an activity or sleep tracker to the app. The first two weeks of recording will help you figure out where you are with respect to your daily routines and what changes you can possibly make to address any issues.

When the pictures come to our server, we put them on a time line to make it easy for us to see when you eat. We call it a *feedogram*. You can look at your own feedogram through the app. See the next page for an example.

After the first 2 weeks of assessing your own circadian habit, it takes 12 weeks to gradually adopt a new habit and have it affect your genes. Our existing habits and environments are like another layer of information affecting our DNA. This is called the *epigenetic code,* and it is so powerful that it reinforces our habits in a way that makes us feel like we cannot escape them. When you try to change your habits—whether it is exercise, a new diet, or even a new eating routine, the old epigenetic code is what makes it difficult to make the change a new habit. This is where you need some willpower to fight against your old habits and make new ones. As your body sees the positive results these new habits bring, it will

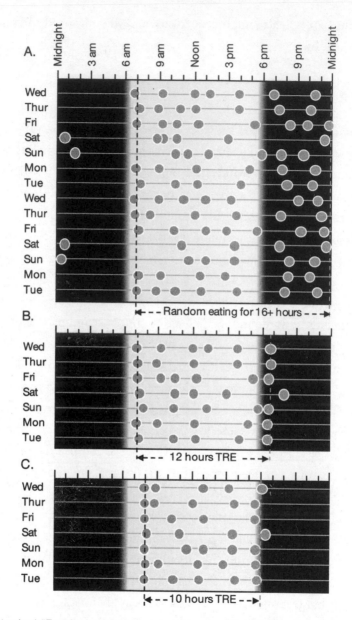

A typical "Feedogram" of one person who (A) used to eat randomly from 6:00 a.m. until midnight, (B) after adopting a 12-hour TRE for one week and (C) following a 10-hour TRE for one week. Each horizontal line represents one 24-hour day, and the position of each circle represents the food/beverage the person ate at the respective time.

slowly get used to them and your old epigenetic codes will be replaced with new codes that will automatically nudge you to stay on your new routine.

So far we have thousands of participants, and we've used their data to make some very important findings. Many of the stories included throughout the rest of the book come from people who have participated directly in our studies or from people who are following the program and have contacted me directly. One thing we learned almost immediately was that people eat far more frequently than they realize. For instance, most people think they eat about three times a day, but a full third of our participants eat almost eight times a day, and they eat well into the night.

As you follow the program, you can also make another big contribution to science by recording your new habits and telling us what went right and what went wrong. This will help us guide you, and your experience will also help others.

PART II

The Circadian Lifestyle

CHAPTER 4

A Circadian Code for the Best Night's Sleep

Now that we know how our circadian code works, the next step is to hack it. We want to get the most out of our daytime activities and nighttime rest. The goal is twofold: First, we want to adjust our activities to the optimal times of the day that are most in sync with our clocks. We want to eat when we metabolize foods the most efficiently, we want to be active when our brain and body are functioning at their peak levels, and we want to get the right amount of sleep so that we can do it all again tomorrow. Second, we can fix disruptions and retrain our clocks to improve our health.

It would be a totally rational guess to assume that the first thing we need to fix is our eating pattern. But in reality, circadian clocks will readjust best when addressing our evening activities, namely limiting our access to light and enhancing sleep. The reason is because sleep is not a passive experience: The human body begins to get ready for the day the night before. Just like we start off the New Year with a celebration on December 31, sleep is the beginning of our biological day, not the end.

Every day, our body battles with lots of stressors that create cellular damage. At night, we aren't just making necessary repairs to the body; the brain is also busy consolidating memories and sending out instructions to prepare us for the next round of activity. The changes that happen at night are absolutely critical to how we feel the next day. That's why when we are in good health and have the right amount of sleep, we wake up feeling refreshed.

The Stages of Sleep

Great sleep is created when there are cycles of quiet sleep and active sleep. The quiet sleep takes place in three stages that occur in a specific sequence: N1 (drowsiness), N2 (light sleep), and N3 (deep sleep). Unless something disturbs the process, you will proceed smoothly from one stage to the next, and as you do, your body and brain perform different functions depending on your clocks. First, in making the transition from wakefulness into light sleep, you spend only a few minutes in stage N1 sleep, but your body and brain change rapidly: Your body temperature begins to drop, your muscles relax, and your eyes move slowly from side to side. During stage N1 sleep, you begin to lose awareness of your surroundings, but you can be easily jarred awake.

The N2 stage, or light sleep, is really the first phase of true sleep. During this stage, your eyes are still and your heart rate and breathing slow down. Brief bursts of brain activity called *sleep spindles* occur, as brain waves speed up for roughly half a second or longer. Some researchers believe that sleep spindles play a role in consolidating memories.

Stage N3, or deep sleep, occurs as the brain becomes less responsive to external stimuli, making it difficult to wake up. Your breathing becomes more regular. Your blood pressure falls, and your pulse rate slows 20 to 30 percent below the waking rate. Your blood flow is directed less toward your brain, which cools measurably. Right before this stage ends, the muscles that allow you to be upright against gravity become paralyzed, which prevents you from acting out your dreams. However, there are some real sleep disorders—like sleepwalking and sleep eating—in which this change doesn't occur. A loss of sleep during this stage may play a role in reducing daytime creativity, mood, and fine motor skills.

These three stages of quiet sleep alternate with periods of active sleep, which is referred to as *REM sleep*, which stands for *rapid eye movement sleep*. During this time, your body is still but your mind is racing. Your eyes dart back and forth behind closed lids. Your blood pressure increases, and your heart rate and breathing speed up to daytime levels.

Dreaming also occurs during REM sleep. We typically have between three to five cycles of REM sleep per night, occurring every 90 to 120 minutes. The first episode usually lasts for only a few minutes, then REM time increases progressively over the course of the night. During this time, the brain focuses on learning and memory.

Each time you move from quiet sleep to REM sleep, you complete a sleep cycle. For optimal health, you need a balance of the different types of sleep throughout the night. Adults need at least 7 consecutive hours of sleep each night. So, if you short yourself by 90 minutes or more, you lose the equivalent of one entire sleep cycle. When you sacrifice a cycle or more of REM sleep, your circadian rhythm may be disrupted.

Within that 7-hour period, there is a critical 4-hour window. You may notice that between the hours of 10:00 p.m. and 2:00 a.m., or in the first 4 hours after falling asleep, you get some of your best sleep. This is because these first few hours go toward paying back your sleep debt. They neutralize the urge to sleep or the tiredness you feel before going to bed. This is why it may be harder to go back to sleep if you wake up after that 4-hour period: You no longer have the sleep debt that was making

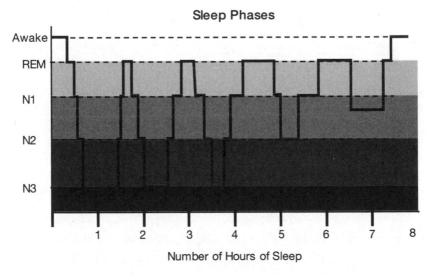

The different stages of sleep during one 8-hour night in bed.

you tired in the first place. The next 3-plus hours of sleep go toward nurturing your brain and body, giving it the additional time it needs for repair and rejuvenation.

Shift workers who have to sleep during the daytime also experience circadian rhythm disruption. Since this is not the typical time nor the optimal lighting for the circadian clock to send its signal for sleep, shift workers aren't able to get the maximum number of sleep cycles during daytime, even if they try to sleep for 7 hours. This is why when you nap in the daytime you can rarely sleep for more than 2 to 3 hours: Your circadian code won't allow it.

Understanding Sleep Debt

As soon as we wake up, our SCN clock begins keeping track of wakeful time. For every hour we stay awake, we later have to sleep 20 to 30 minutes. In the evening, the organs' unique clocks synchronize with one another to create the perfect condition for sleep. The pineal gland inside the brain begins to produce the sleep hormone melatonin. At the same time, the heart clock instructs your heart rate to slow down, and the SCN instructs the body to cool down. Then, when the timing is right and the lights are low, you go to sleep.

Every night adults should give themselves 8 consecutive hours of sleep opportunity, and children should have 10 hours of sleep opportunity. That includes getting into bed, settling down, and then falling asleep. Children should be sleeping for at least 9 hours a night; adults should sleep for no fewer than 7 hours.[1,2]

Sleep debt is the difference between the amount of sleep you should be getting and the amount you actually get. So, if you slept for 6½ hours last night, you're beginning your day with 30 minutes of sleep debt. When you go to sleep the following night, you first repay this debt from the previous night. That means even if you sleep 7 hours the second night, it only counts as sleeping for 6½ again. That's one of the reasons why we often sleep late on weekends: It's the body's way of repaying your entire debt.

Sleep debt increases our propensity to sleep, while circadian rhythm instructs when we should sleep. For instance, if you are awake for 2 days, you have too much sleep debt to unload in a single night. You'll go to sleep, but your circadian clock won't allow you to sleep continuously for 16 hours. The first night you might sleep for 8, 9, or maybe 10 hours, and then the circadian drive to wake up kicks in. The next day you're still sleepy because the clock is telling the brain it's time to stay awake, but the sleep debt is telling the brain that you ought to go back to sleep. That conflict goes on into the following night, again, when you'll sleep a little longer until you catch up.

Napping Counts Toward Repaying Your Sleep Debt

A short nap during the day is one way to repay your sleep debt. For example, if you had a sleep debt from the week of 2 hours and you take a Saturday afternoon nap, it's possible to repay that debt in one nap. But be careful not to sleep too long: Sleep time is a function of your circadian clock and how many hours you are awake that day. A long afternoon nap will dissipate some of the sleep pressure that was building up since morning, but the longer you sleep in the afternoon, the further you may push your nighttime sleep, making it difficult for you to fall asleep when you want to later that night.

The only times when napping really works against you are when you are jet-lagged, if you are a true shift worker and you want to sleep at night, or if you are really trying to move your bedtime to earlier in the evening. In these instances, it's better to build up your propensity to sleep at night, and then reset your clock the next morning.

The U Curve of Sleep and Longevity

There are real benefits to achieving the prescribed number of hours of sleep. From tracking a million individuals, researchers have identified a pattern, known as the *U curve of sleep and longevity*.[3] People who consistently sleep too little are more likely to die early than those who get the full 7 hours of sleep each night. Similarly, people who sleep as much as 10 to 11 hours are also likely to live shorter lives. The majority of people who had an ideal body mass index (BMI), a standard health measurement that tracks a healthy weight-to-height ratio, were also shown to sleep 7 hours a night. The bottom line is that too much or too little sleep can be detrimental.

One way to see if you are in this sweet spot of the U curve is to track your sleep habits. You can use the chart in Chapter 3 to fill in when you go to sleep and wake up, or you can use the myCircadianClock app (available at mycircadianclock.org) or any wearable sleep tracker. The more you know about your sleep patterns, the easier they are to correct. The guidelines on page 75 are ideal for maintaining your circadian code no matter your age.

The Myth of Ancestral Sleep Patterns

There seems to be some Internet myth that ancestral people slept for a few hours; woke up in the middle of the night; did some activity, like sex or eating; and then went back to sleep. However, the research doesn't support this. As recently as 2016, scientists conducted studies on indigenous groups, including the Hadza of Tanzania and the Toba of Argentina, who don't have access to electrical lighting.[4,5] They sleep in their huts, or sometimes even in open fields. When scientists put sleep trackers on these people for many days at a time, they didn't find any sign of two-phase sleep. These people slept for the typical 7, 8, or 9 hours. They would go to bed around 9:00 or 10:00 p.m. and wake up at the crack of dawn.

Two-phase sleep is actually a more common pattern in our modern lifestyle. Lots of people wake up after 3 to 4 hours (that window when sleep debt is paid off) and find it very hard to fall back to sleep. Frustrated, they might start working on the computer, start reading a book, or go to the kitchen to get a bowl of cereal. This type of sleep works in opposition to your circadian code and is one of the habits this book will help break.

Recommended Hours of Sleep Across the Lifespan

	Age	How many hours of sleep		Minutes in bed before you fall sleep			No. of times waking up for >5min		
		Ideal	Not recommended	Normal	Borderline	Talk to your doctor	Normal	Borderline	Talk to your doctor
Newborns	0–3 months	14–17	<11 or >19	0–30min	30–45 min	>45 min	Normal to wake up a few times		
Infants	4–11 months	12–15	<10 or >18	0–30min	30–45 min	>45 min	Normal to wake up a few times		
Toddlers	1–2 years	11–14	<9 or >17	0–30 min	30–45 min	>45 min	1	2–3	>4
Pre-schoolers	3–5 years	10–13	<8 or >16	0–30min	30–45 min	>45 min	1	2–3	>4
School-aged children	6–13 years	9–11	<7 or >15	0–30 min	30–45 min	>45 min	1	2–3	>4
Teenagers	14–17 years	8–10	<7 or >13	0–30 min	30–45 min	>45min	1	2	>3
Young adults	18–25 years	7–9	<6 or >11	0–30 min	30–45min	>45min	1	2–3	>4
Adults	26–64 years	7–9	<6 or >10	0–30 min	30–45 min	>45 min	1	2–3	>4
Older adults	>65 years	7–8	<6 or >10	0–30min	30–60min	>60min	2	3	>4

M. Ohayon et al., "National Sleep Foundation's Sleep Quality Recommendations: First Report," Seep Health 3, No. 1 (2017): 6–19.

Are You Sleeping Well?

Ask yourself the following three questions to get a clear picture of your sleep quality.

QUESTION #1: *When do you go to bed, and how long does it take for you to fall asleep?*

First of all, let's lower the bar a bit: Most people do not shut off the lights and fall immediately asleep. For an average person who has good sleep habits, you should be able to fall asleep within 20 minutes of getting into bed and shutting off the lights. During this 20 minutes there should be nothing else between you and sleep. No book. No phone. No light.

If you struggle to fall asleep and you're in bed for more than half an

---

Spending the Day Outside Makes Nighttime Indoor Light More Tolerable

When you spend a full day (4 to 5 hours) at the beach or at a park with bright daylight, you're less sensitive to the effect of bright indoor lights at night. When I was camping in the Maasai Mara National Reserve in Kenya, I was exposed to at least 8 hours of bright light every day, and no artificial light. It was as if I was living in a world 1,000 years ago, and each night I slept well: 7½ hours a night. A week later I was working in a lab in Nairobi, where there were lots of windows that brought in natural light. It was as if I was spending 3 hours a day in natural light, even though I was indoors most of the time. Those nights I slept well, but not as long as on the nights I spent camping. Then I returned to San Diego. My office there has limited natural light, and I barely had an hour of daylight exposure. Look at my sleep charts on page 77. You can see that back at home, I wasn't sleeping well due to lack of sunlight during the day.

Pay attention to how much light you get during the day. If the only time you can recall seeing the sky is when you are driving to or from work, chances are you are not getting enough natural daylight. Try to walk outdoors even for a few minutes during a break in your day. Better still, have meetings outdoors or next to a large glass window where you can catch your fair share of bright daylight.

hour, turning and tossing, that's a sign that you have difficulty getting into sleep. This is the definition of insomnia: difficulty falling asleep.

The main culprits for insomnia are:

- Worry: increases the stress hormone cortisol, which is meant to keep us awake

- Too much food: keeps core body temperature too high for sleep

- Too little physical activity: reduces the production of the muscle hormone that promotes sleep

- Too much time spent in bright light in the evening: activates melanopsin and reduces melatonin production

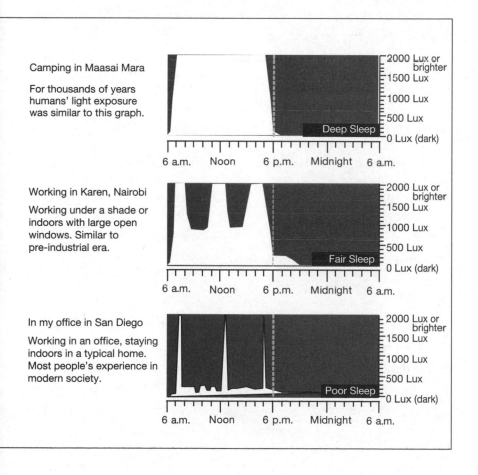

QUESTION #2: *How many times do you wake up during the night?*
Fragmented sleep is defined by waking up more than once during the night for at least a few minutes, to the point where it's difficult to go back to sleep. This type of sleep is not optimal, because the brain registers only the time you sleep, and it responds as if it isn't getting any sleep at all during these periods of fragmented sleep. For instance, if you were in bed for 8 hours but woke up three or four times, then your brain might register only 4 to 5 hours of actual sleep. Even if you woke up for only 10 to 15 minutes each time, it takes additional time to get back to that deep sleep phase, and you miss out on this continuous uninterrupted sleep.

As we age, our sleep becomes more fragile and it is very common to experience fragmented sleep. Our arousal threshold decreases with age, so we wake up to simple noises or disturbances. However, it is possible to sleep through the night without fragmented sleep, especially if you align your sleep time to your circadian code.

The main causes of fragmented sleep are:

- Dehydration

- Ambient temperature being too hot or cold

- Acid reflux caused by eating too late in the evening

- Sleeping with a pet

- Snoring/sleep apnea

- Other noise

QUESTION #3: *Do you feel rested when you wake up in the morning?*
If you need to wake up to an alarm clock, or if you wake up feeling sleepy or foggy, then you aren't waking up feeling rested, and it's likely that you did not get enough sleep.

Poor Sleep Disrupts Your Circadian Code

Fragmented sleep and insufficient sleep affect almost one in three adults in the United States. That means in the morning when you're driving to work, look to your left, to your right, and in front of you: One of those drivers is actually sleep-deprived (if it isn't you).

No matter how old you are, poor sleep leads to poor performance, which has both short- and long-term consequences. In the short term, after only one night of poor sleep, adults may experience brain fog and confused thinking the next day that affects their decision making, reaction time, and attention. For instance, we know that when you're sleep-deprived, your performance is worse than somebody who has had two alcoholic drinks.[6,7] It goes without saying that children and teens who get less sleep don't perform as well at school as those who get more sleep. Even young children are affected, and lack of sleep can make them seem more irritable or difficult.

When poor sleep becomes habitual, long-term consequences are more serious. One study showed that children with attention deficit hyperactivity disorder (ADHD) have fewer symptoms when they have had enough sleep at night and exposure to light during the daytime.[8] Adults with poor sleep habits are more likely to develop anxiety and depression, and seniors may experience memory impairment.[9,10]

Sleeping less also means doing something else more, and more often than not, it means exposure to additional light at night, or to more food during the day or night. Sleep deprivation directly affects our hunger and satiety hormones, like ghrelin and leptin, both of which have a circadian nature. Ghrelin is produced in the stomach whenever the stomach is empty, and it is the signal to the brain to feel hunger. Leptin is produced in fat cells and signals the brain that you are full. However, poor sleep patterns disrupt these signals and make us more prone to overeat because the brain isn't getting either of these two messages. One UK study tracked hundreds of children from ages 3 to 11. They found that children who were given dinner at the same time and put to sleep early in the evening

Signs of a Chronic Sleep Problem

Waking up with joint pain may be a sign that you did not get a good night's sleep for several days in a row. Inflammation in the body is supposed to reduce during sleep. If you don't sleep long enough, inflammation doesn't have time to subside. You might find that if you sleep less than 6 hours a night for 3 or 4 nights, when you wake up your joints are stiff, or you might have a pain in your knee. However, if you can get on a better sleep schedule, you may find that the pains go away without any medication, without any exercise, and without changing your diet.

every night were far less likely to develop obesity by age 11.[11] These children had a strong circadian code for sleep, as well as for metabolism.

Unfortunately, having a late-night snack is the general bedtime routine for many people. In the same way we know that leptin and ghrelin disrupt circadian eating patterns, there are many mechanisms that cause our body to overeat when we don't have enough sleep. We think the reason is the brain wants to ensure that we have enough calories to cover activity during those hours. But in controlled studies in Ken Wright's sleep lab, participants who reduced their sleep from 8 hours to 5 hours consistently overate more calories than what would be required to fuel

Light at Night Enables Bad Habits

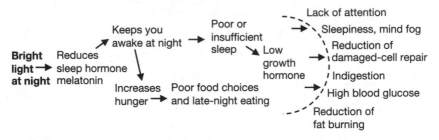

Bright light at night has a bad domino effect on both the brain and the body. If you can manage your exposure to light and fight the temptation to eat at night you can break this pattern.

a few extra hours of wakefulness.[12] This finding does not mean that the brain actually needs additional food to function properly without sleep. Rather, a sleep-deprived brain, or one exposed to bright light at night, craves excessive calories that it does not need, resulting in weight gain.

In fact, even a sleepy brain may work better without adding food to the mix for hours at a time. Research in Mark Mattson's lab at the National Institutes of Health has shown that mice that undergo a longer fasting period have better brain function, as keeping to a restricted eating time strengthens the connections or synapses between brain cells.[13] A stronger connection between neurons means the brain can think better and remember better, regardless of how rested we are.

Let's Get to Sleep!

The basic lesson for improving sleep is to increase the drive to sleep in the first place and avoid the factors that suppress or disrupt sleep.

In the daytime, the drive to sleep is affected by many factors:

- *Length of time one has been awake:* Sleep drive increases with every hour we are awake. If you want to go to bed early, you should wake up early as well.

- *Exercise or physical activity:* Physical activity, particularly outdoor activity under the sun or under diffuse daylight, increases the drive to sleep.

- *Timing of caffeine intake:* Caffeine reduces our sleep drive and keeps us awake. Reducing caffeine after midday is a good general rule of thumb.

Food, Timing, and Sleep

Eating late at night is not only bad for metabolism, it also affects sleep. This habit interferes with both falling asleep and maintaining deep sleep. In order to fall asleep, our core body temperature must cool down by almost 1°F. But when we eat, our core body temperature actually goes up as blood rushes to the gut (the core) to help digest and absorb nutrients. So, eating late at night prevents us from getting into a deep sleep. To have a good night's sleep, we should have our last meal at least 2 to 4 hours before going to bed to ensure that the body is able to cool down.

In my lab, we have found that mice that follow a time-restricted eating (TRE) schedule of 8 or 9 hours also sleep better. Their core body temperature falls and they get into deeper sleep. Surprisingly they don't sleep longer than the mice that eat randomly, but electrical recording of their brain shows that their sleep is deeper and perhaps restful (we cannot ask a mouse if it slept well). We don't know the reasons, but we often find time-restricted eating improves sleep by increasing the arousal threshold. In other words, the basic sleep drive does not change under TRE, but TRE puts the mice into deeper sleep (high arousal threshold).

Through our app, myCircadianClock, we have observed many people who do a 10-hour TRE report substantial improvement in sleep. In fact, some of them do TRE not for weight loss but to sleep better at night.[14]

Drinking alcohol before bed has a very different effect even compared to eating sweets, but it is just as disruptive. In an oxymoronic way, alcoholic beverages dehydrate you, and the more you drink before bedtime, the thirstier you will be in the middle of the night. That hangover is partly your brain reacting to not having enough fluids. So, while some may use a late-night drink to fall asleep, they generally have difficulty staying asleep. If you like a cocktail after dinner, a better habit would be to have your drinks before dinner, or during your dinner, provided it is 2 to 4 hours before bedtime.

Once you are used to having a good night's sleep, more than a glass of wine will seem less appealing. For instance, one man in our study group used to have three or four cocktails after dinner. Once he started cutting down on his alcohol, he found that he was sleeping better. After a while, he gave up his cocktails entirely. He told me, "I don't enjoy my cocktail anymore, and I really like my better sleep at night."

Jay Couldn't Sleep through the Night

Jay, a 41-year-old with a high-stress management job, was so busy at work that he never had time to exercise. When I first met him, he was at least 40 pounds overweight for his small frame. He told me that his sleep was terrible; he woke up two or three times a night, and then started immediately worrying that he wouldn't fall back asleep. To compensate, he tried to be in bed for 8 hours every night, but because his sleep was fragmented, he never woke up feeling refreshed.

I suggested that he try TRE to consolidate his meals into 10 hours each day. A few weeks later, Jay contacted me and he sounded great. It took him very little time to adjust to this eating routine, even though he had previously tracked that he was eating up to 15 hours a day. He reported that in just a few weeks he was able to sleep through the night: 7 continuous hours. He had also lost around 10 pounds, which was no surprise to me. But what

---

The Acid Reflux Problem

Some people wake up in the middle of the night with acid reflux, or a bad feeling in their stomach that only food seems to soothe. The food of choice seems to be a bowl of cereal, which poses two problems: The protein in milk triggers the stomach to produce more acid, and the carbohydrates in cereal cause a spike in blood glucose.

Consult with your doctor if this is a persistent problem for you, but my advice is to stay away from food as a cure and take your acid reflux medication instead. But more important, stay away from that late-night snack, which is actually what is causing the reflux in the first place (you'll learn all about this in Chapter 9). When you stay away from late-night snacks and eat healthier and earlier, you may be able to get off of your medications, because your acid reflux will reduce to an extent where you don't need drugs and you don't need to wake up to eat.

was really interesting was that Jay told me he felt refreshed each morning and was more productive after only 7 hours of sleep. He didn't have to stay in bed for 8 hours, and he was able to devote that extra hour to his family.

Sleep Is Inhibited by Light at Night

The easiest sleep fix is to maintain a dark sleeping environment. Everyone knows it is hard to fall asleep under bright light. Your circadian clock prevents this. The blue light sensors pick up the bright light to suppress sleep and promote wakefulness. However, other colors in the light spectrum, specifically orange and red, are less effective at suppressing sleep.

Pay attention to the type of light that you are exposed to in the evening. We cannot go back to the dark ages or turn off all lights after sunset, but managing our exposure to light can have a huge impact on improving sleep and maintaining health. If you find that you're particularly sensitive to light, try an eye mask. Make sure it's comfortable and stays in place if you move. If it's too tight, your ears may feel sore in the morning, but the right one can really improve your sleep.

Teens and Sleep

Teens are especially susceptible to light and breaking the circadian code. Not only are they more likely to stay awake in the evening because of homework or their activities, there are also studies showing that teens are very sensitive to light.[15] That means exposure to bright light in the evening delays their sleep and lowers their melatonin production.

We can do at least two things to help our teens. First, we can prepare an early dinner in the evening so that they have an empty stomach before they go to sleep. They are most likely to fall asleep 3 to 4 hours after dinner. At the same time, we should also educate them by telling them about the importance of darkness and sleep. And perhaps we can establish a sleep-friendly environment for them to do their homework, including a table with a spotlight or lamp that illuminates the table but not their eyes.

Small Lighting Changes Can Have a Huge Impact

I'm not suggesting we spend the evenings in a dark room until we go to bed. There are many techniques and products that can help reduce our exposure to blue light. For instance, in the evenings, shut off overhead lights and use table lamps instead. For rooms like the kitchen and bathroom, dimmer switches will help you easily reduce ambient overhead light. There are even lights that can be programmed to turn off and on at different times of the day. These strategies benefit teens as well as adults, as it is an easy fix to reduce the amount of light we have at home. You'll learn more about specific lighting products and techniques in Chapter 8.

Michael Gorman at the University of California, San Diego, did a simple experiment with mice and light.[16] He turned on a very dim light in the mouse house at night. It was dimmer than a typical night lamp many people use in their home and was almost equivalent to dim lights coming out of indicators on our TV, phone, or similar instruments. Surprisingly, mice were very sensitive to even this level of dim light, and their sleep cycle was affected. New research coming out of Samer Hattar's lab at the National Institute of Mental Health indicates that even dim light from innocuous sources can compromise sleep and circadian rhythm in animals. While this is yet to be tested rigorously in humans, anecdotally we have seen that many people are very sensitive to dim light, and they sleep better with an eye mask or in a completely dark room in which every possible source of light is covered.

If you wake up in the middle of the night to get a drink of water or go to the bathroom, turning on a light makes it infinitely harder to fall back asleep. Try to keep light to a minimum. Keep a glass of water next to your bed, which will save you the trip in the first place. Or, if you need to use the bathroom, here's where sleeping near your cell phone actually comes in handy: Use the flashlight feature to light up the floor to help you find your way.

I always have a glass of water next to my bed. Some people think that if they drink water in the middle of the night, then they're going to wake up again. In reality, you're not drinking more than a few ounces. In fact, ignoring your thirst is far worse: A dry throat is the reason you probably woke up in the first place.

Hacking Your Way to a Better Night's Sleep

A good night's sleep ensures better performance the next day. It puts you in better alignment with your circadian code by increasing growth hormone production while you rest, rejuvenating your brain and body. It increases your cortisol production in the morning, which helps with alertness, and balances your hunger and satiety hormones for stronger, more efficient metabolism. Best of all, it synchronizes all of your internal clocks so that your whole body is working at peak performance.

If you are consistently not getting a good night's sleep, or if you are waking up at night, try the following techniques.

Turn Down the Temperature

The body has to cool down during nighttime to sleep. It's a good idea to reduce the temperature in your bedroom to 70˚F or lower so that your skin feels cooler. When this happens, blood flows toward your skin to keep your skin warm. Since the blood is flowing away from the core of the body, the core body temperature can fall and you will fall asleep much easier.

If you cannot control the thermostat in your home, take a shower or a warm bath just before going to bed. Warm water also forces blood flow toward the skin and away from the core.

Some people fall asleep, but after a few hours, they wake up feeling too hot. Experiment with your blankets to find what works best for you. If the blanket is not the culprit, think about your mattress. Foam mattresses are known to capture heat. In the first few hours, the mattress actually helps you cool down, but after a few hours, foam mattresses can reflect heat back to your body and warm you up.

Turn the Sound Up or Down

In many cities, sounds and sirens make it difficult to fall asleep. Triple-pane glass in your windows will cut out sound to a great extent. For years, shift workers have adapted with the trick of having a fan blowing in their bedroom so that the humming noise suppresses or blocks out all the other tiny noises that might disturb them. A more modern approach would be a white-noise machine (or white-noise app). These devices can make it easier to fall asleep and stay asleep by fighting noise with noise: The machine creates a wall-of-sound defense, protecting you from intruding noises that could engage your brain during sleep.

Some people actually find sound soothing, helping them to fall asleep. Put your radio or smartphone on a timer and play relaxing music at a low volume for a few minutes as you're falling asleep.

For some people, like me, even small noises (like a noisy air conditioner or a partner who snores) can wake them up. This is where earplugs come in. When I travel, I always bring earplugs. Yet not all earplugs are created equal. Some are soft. Some are hard. Some are silicone. Some are like a sponge. You may have to try a few to find the one that's most comfortable. See which one fits your ear well, so that you don't have a sore ear canal in the morning. Once you find the right earplugs, you will experience a better night's sleep immediately: They make a huge difference.

Age Isn't an Excuse: Better Sleep Is Available to Everyone

We aren't programmed to sleep less as we get older. It's just that with age, we become more sensitive to different factors that wake us up. Try the tips suggested in this chapter, because they do make it easier to fall asleep. For instance, I used to sleep only 6 hours. After I adopted all these tricks, now I can easily sleep for 7, 7½, or sometimes even 8 hours, even when I'm traveling.

Is Snoring Affecting Your Sleep?

Snoring may be the butt of many sitcom jokes, but it is no laughing matter. Adults may snore when we put on some extra fat or when our muscle becomes weak around the breathing canal. In both of these scenarios, our wind pipe gets obstructed when we are asleep, and that leads to snoring.

Snoring is rare among children, but it can occur when they are congested because of sickness or allergies. Both children and adults who have stuffy noses will breathe through their mouth at night, and tend to snore. Mouth breathing reduces the amount of oxygen that goes into the brain. That also puts the brain in a hypoxic, or low-oxygen, state, which can increase the chance of getting dementia and various brain-related problems like memory loss.

Easy Tips to Stop Snoring:

The easiest, least invasive fix for snoring is to try a gentle saline spray or neti pot. These are meant to clean and open up stuffy noses. Saline spray is safe for both adults and children to use every day.

The second simple thing you can do is use a sleeping aid that keeps your nose open. There are two main types: ones that pull the skin on your nose open, like a Breathe Right nasal strip, and ones that are inserted inside your nose to open the airway. Not only do these keep the nose open throughout the night, they also allow you to breathe in more oxygen, which will improve the quality of your sleep quite a lot. Some days, if I'm tired at the end of a long day of work, I actually put one of those Breathe Right strips on my nose when I'm driving back from work because I know that one aspect of feeling tired is my brain didn't get enough oxygen during the daytime (because I'm chronically stuffy). In that 30-minute commute, I get enough oxygen so by the time I hit home, I'm really full of energy again.

If snoring continues even after these over-the-counter fixes, see an ear, nose, and throat specialist (ENT) or a pulmonary medicine sleep specialist.

Sleep Apnea Is Serious

Obstructive sleep apnea (OSA) is one of the major causes of sleep deprivation. It occurs when you have a blockage or obstruction in the nasal cavity or throat or have a floppy tongue that obstructs your airways either partially or completely during the night. The obstructions deprive the brain and the body of oxygen and cause an automatic response that wakes you up just enough so that you breathe again, although you may not wake to the point of consciousness. These upsets can occur all night long, yet people with sleep apnea very often have no clue. Instead, they'll wake up in the morning without feeling refreshed. Other subtle clues include waking up with a dry mouth or having to repeatedly use the bathroom in the middle of the night.

Some people with sleep apnea snore, but not all of them. And not all snoring is considered to be sleep apnea. Your partner might be a better detective than you in determining if you are suffering from sleep apnea: If you have been told that you hold your breath during the night, you may be affected.

Sleep apnea affects not only the quality and quantity of your sleep but also your brain health. Cognitive problems, such as deficits in memory, attention, and visual abilities, frequently accompany OSA. It is also a major risk factor for heart disease and stroke since as many as two-thirds of people with underlying sleep apnea have high blood pressure.[17]

A sleep study can help you determine if you are suffering from sleep apnea. The standard treatment for sleep apnea is a device referred to as a CPAP (continuous positive airway pressure) that is prescribed by a doctor; trained medical staff guides you on how to use the machine. It is a mask that you wear over your nose and mouth that is hooked up to a machine to make sure that you have a constant supply of air. There are other devices and apps that can monitor your oxygen intake as well.

Sleep Medications

While effective, sleep medications have *never* been tested for continuous use for more than 6 months. We don't know what the long-term benefits or adverse side effects are of most of these medications. Please keep that in mind if you have been tempted to ask your doctor for a prescription.

Sleep medications fall into two different categories. The first are the ones that improve your ability to fall asleep, like Ambien (zolpidem), Lunesta (eszopiclone), and Restoril (temazepam). If you fall into the camp who needs this type of drug, consider trying melatonin supplements first, as they reduce the time between going to bed and falling asleep.[18]

The second type of medication is for people who cannot stay asleep, or who wake up too many times throughout the night. These sleep medications, like Silenor (doxepin), help people with fragmented sleep get uninterrupted sleep, but some are so strong that in the morning people still experience sleepiness and brain fog. These medications help you fall asleep, but they don't help you wake up.

Sleep medications are not a permanent cure for your sleep problems; when you get used to them, your brain relies on the medication to help you fall asleep. If you are a frequent user, or have been taking sleep medication for a long time, it can take up to 2 weeks to try to fall asleep without them. And sleep medications have lots of adverse side effects, including dizziness, light-headedness, headache, gastrointestinal problems, prolonged daytime drowsiness, severe allergic reactions, and daytime memory and performance problems. What's more, there are no longitudinal (long-term) studies that show the effectiveness of sleep medications for more than 6 months.

My advice is that if you really think you need sleep meds, try a high-quality melatonin supplement first.

Melatonin Supplementation

The sleep-promoting effect of melatonin supplementation has been known for almost 5 decades. We need melatonin to sleep. The body

Preparing for Air Travel

When you're flying at 30,000 feet, the plane is pressurized to only 15,000 feet. That means you're actually spending that flight time on top of a mountain that is 15,000 feet high. It's no wonder you get a headache, your brain becomes foggy, your breath light, and you cannot sleep on a plane: The lack of oxygen presents a problem. This is where, again, a breathing aid comes in really handy, as it opens the nostrils, allowing us to breathe at least 20 to 50 percent more air (and more oxygen) than the person next to us. That reduces plane fatigue and jet lag when we reach our destination.

Think of your flight time as your best opportunity to sleep. Instead of watching TV, put your eye mask on, your earplugs in, and try to sleep. While you're at it, skip the airplane food if it is served at a time that is not aligned with your normal eating pattern: It's not necessarily healthy for your circadian code, and it certainly isn't going to put you to sleep.

produces its own supply, but as we age, the pineal gland produces less melatonin at night. A 60-year-old produces one-half to one-third of the melatonin of a 10-year-old. Therefore, boosting nightly melatonin with a pill may be reasonable if you are having problems getting to sleep.

Try taking melatonin supplements 2 to 3 hours before going to bed. However, be aware that melatonin can interfere with blood glucose regulation. Blood glucose naturally goes up after a meal and takes an hour or longer to come back to its normal level. Taking melatonin after a meal slows down the decline in blood glucose to the normal level. Therefore, it is a bad idea to take melatonin right after eating: Wait for at least an hour or two after a meal so that the melatonin doesn't interfere with your blood glucose level.

In many individuals, natural melatonin levels begin to rise 2 to 4 hours before their most frequent bedtime. If this is true for you, the best time to take melatonin is 2 hours before bedtime. This means that if you are planning on going to bed around 10:00 p.m., take your dinner at 6:00 and your melatonin at 8:00.

The effective dosing of melatonin seems to vary from person to person. Some people are very sensitive and a small dose of 1 milligram may be more than enough, while other people take 5 milligrams to get better sleep.

Behavioral Techniques for Better Sleep

1. Don't look at your watch/clock/phone when you cannot get to sleep or if you wake up in the middle of the night, because the light from these devices will trigger your melanopsin. It really doesn't matter what time it is when you wake up in the middle of the night, and there's no benefit to starting to worry about not getting enough sleep. If you need an alarm to wake up at a certain time, that's fine: Set it and cover it so that even those lights don't disturb your sleep.

2. Don't create stress around bedtime or worry that you will wake up late the next day. That's what alarm clocks are for. Relying on alarm clocks is not ideal, but as you are working on improving your circadian code, there is a place for them in your life. Instead of worrying that you won't wake up on time, try deep belly breathing to relax your body and mind.

3. Don't create stress about your last night's sleep and worry that you'll have the same bad experience again. You are in control of your sleep. By following the recommendations we've laid out in this chapter, it's quite likely that your sleep will improve, bit by bit, every night.

4. Don't create stress about the number of hours you're currently sleeping. If you are feeling fine and restored the next day, you may not need as much sleep as others. But if you don't feel rested and refreshed in the morning, or if you feel sleepy during the late afternoon, try some of the tips in this chapter.

5. Don't use the bedroom for anything other than sleep. It's not a study or a living room or a home theater.

The Best Ways to Wake Up

Is there any room for improvement to optimize waking up?

- The best way to wake up refreshed is to have enough sleep by going to bed early.

- Get some bright light immediately after waking up. Open your curtains or turn on your overhead light. Get as close to the window as possible.

- Take a quick, 5- to 15-minute morning walk. Check your plants, check the bird feeder, play with your dog in the backyard, brush off your car. Do anything that will take you out of the house and into bright daylight.

- Try to be consistent and wake up at the same time every day. If you are waking up 2 hours later on the weekends, it is a fair sign that you are not getting restorative sleep during the week.

CHAPTER 5

Time-Restricted Eating: Set Your Clock for Weight Loss

All of nutrition science is based on two experiments. The first proved the notion of calorie restriction: If we eat less, we'll lose weight and achieve better health. This experiment was done in the early part of the 20th century, and ever since, people have been counting their calories.[1,2]

The second experiment (in fact, there are more than 11,000 studies using this model) supports the notion of a "healthy diet." In this experiment, a pair of genetically identical mice were fed two different diets, one with a healthy balance of carbohydrates, simple sugars, proteins, and fats, and one high in fat and sugar. After a few weeks (which is equivalent to a few months or years for humans), the mice eating the high-fat/high-sugar diet became obese, almost diabetic, and had high levels of fat in their blood and dangerous levels of cholesterol. This finding drives home the notion that the quality of your food—its nutritional content—matters significantly when it comes to your health.

Variations of the same experiment continue to be studied with different macronutrients (proteins, carbs, or fats) and micronutrients (antioxidants, vitamins, minerals, etc.). This research is what drives our current "eat this, don't eat that" thinking. Yet none of this research has proved conclusively that one type of food is best for everyone. It turns out that what's best for you is a balanced combination of various macronutrients and micronutrients in quantities that are large enough to keep you satisfied but not gain weight. But the definition of what is "balanced" is highly contested, as what is optimum for an athlete, an expectant mother, a teenager, a bodybuilder, and a patient with diabetes may be vastly different.

We already knew that mice that don't have a normal clock are predisposed to obesity, diabetes, and many of the same diseases that generally occur in mice when they are given a high-fat diet. What's more, a poor diet is all it takes to break a mouse's hunger and satiety clock.[3] Unhealthy mice continue to eat late into their bedtime and also wake up in the middle of their sleep to snack. We wanted to know how much of their disease was due to the poor diet and how much was due to poor eating habits. So, in 2012, we asked a very simple question: "How much of disease is due to a poor diet, and how much is due to random eating?"

Our experiment with genetically identical mice focused solely on the restriction of time for eating, and it produced amazing outcomes, establishing the idea that it's not only how much we eat and what we eat, but *when* we eat that matters, especially for long-term positive health outcomes. We took pairs of genetically identical mice born to the same parents and raised in the same home and gave one group access to a high fat-diet whenever they wanted. The other group had the same amount of food, but they had to eat all their food within an 8-hour window. The mice with the smaller food window quickly learned to eat the same number of calories as the mice that had access to food all the time. In other words, mice on a 24-7 schedule ate small meals spread throughout the day and night, while mice on an 8-hour schedule ate the same number of calories, just in larger meals within the 8 hours.

What's more, over the first 12 weeks of the study, when the mice ate the same number of calories following the same high-fat/high-sugar diet that in 11,000 other publications had been shown to cause severe metabolic diseases, but within an 8-hour window, they were completely protected from the diseases normally seen with a poor diet. The time-restricted eating mice didn't gain excess weight, and they had normal blood sugar and normal cholesterol levels. We believe that a shortened feeding period provides the digestive system the right amount of time to perform its function uninterrupted by a new influx of food, and enough time to repair and rejuvenate, supporting the growth of healthy bacteria in the gut. This restricted feeding period is in alignment with the mice's natural circadian code, which is why they lost weight and stayed healthier. The benefits continued week after week for an entire year (which is like several years of human life) as long as the mice stayed on the new eating schedule. In fact, the health benefits were far greater than the effect of a drug to treat the same condition. Remember, we did not change the diet and we did not reduce their calories. Timing made the magic.

Later, we ran the same experiment with 9-, 10-, and 12-hour windows and found similar benefits overall. It appears that when mice eat for 15 hours or longer each day, their body reacts as if they are eating constantly. The mice on a 15-hour diet were not very healthy, while those who ate for 8, 9, 10, or 12 hours remained healthy. We systematically examined their health every week, monitoring several hormones and even how their gut microbes changed. We tested which of the 22,000 genes turned on and off at different times of the day in different organs. These experiments spanned many years and were published in peer-reviewed scientific journals.[4,5,6] They have now been replicated in many labs around the world.

Researchers have followed this up with another study, where they combined the initial calorie restriction study with our clock research.[7] They wanted to test if a low-calorie diet was as effective regardless of when it was given. First, they gave the mice their low-calorie food at bedtime, and found there was no weight-loss benefit. But when they gave the mice the same amount of food when they first woke up, the mice lost weight and their eating pattern was aligned with their circadian code.

We have seen similar results in human studies. For instance, a group of Harvard scientists and Spanish weight-loss nutritionists found that individuals who spread their calories over a long period of time—meaning they eat the same number of calories but eat later into the night—did not lose much weight. However, people who ate bigger meals during the day and refrained from eating at night actually lost a substantial amount of weight.[8] This means that regardless of which kind of calorie-restricting diet you follow, when you eat is more important than what type of food you eat.

Stop Eating Like a Shift Worker

In the same way that it's better not to sleep like a shift worker, our experiments show that it's significantly healthier not to eat like one. Our brain clocks are most sensitive to light, but the clocks in our gut, liver, heart, and kidneys respond directly to food. Therefore, just like the first sight of morning light resets the brain clock and tells it that it's morning, the first bite or first sip of coffee of the day tells the clocks in our gut, liver, heart, and kidneys to begin the day. If we change our routine from day to day, our clocks get confused.

In 2015, we did a study looking to see when people actually eat. We told 156 participants to record every meal, snack, and beverage. They used their cell phones and our myCircadianClock app. We found that 50 percent of all participants ate for 15 hours or more every day.[9] That means they ate during almost all of their waking hours. Twenty-five percent of all participants delayed their weekend breakfast by 2 hours compared to how they ate during the week. Even this one breakfast shift disrupted their circadian code; it was as if they were a true shift worker, or living in two different time zones: one during the week and one on the weekend. But what was even more interesting was that when we asked all of the participants when they thought they ate, they responded almost uniformly that they believed they ate within a 12-hour window. They weren't counting their early morning coffee with cream or that last glass of wine or handful of chips or nuts they'd eat after dinner.

Then we asked 10 participants who ate for 14 hours or more and were already overweight (with a BMI over 25, which is the standard measurement) to choose the same 10-hour window each day in which they would eat all their meals. Everything: beverages and snacks included. We didn't give them any instructions on what to eat or how much to eat or how often to eat. They recorded their meals again and submitted them to the app. We gathered the data. And what we found was astonishing: All of the participants lost an average of 4 percent of their total weight in just 4 months. They ate whatever they wanted, and they all lost weight. They also reported that they slept better at night and felt more energetic and less hungry during the day. The benefits of time-restricted eating (TRE) in humans is currently being replicated by other researchers.[10,11,12,13] Clearly, TRE put these people back in sync with their circadian code.

Our findings underscore the importance of one of the primary goals of this program: to align your eating schedule with your circadian code. Start by establishing a 12-hour window for a week or two, and then try to decrease the time you eat by an hour a week. The reason to do this is that the optimum eating window is between 8 and 11 hours. This is because the health benefits that you get from eating within a 12-hour window double at 11 hours, and double again at 10, and so on, until you reach an 8-hour window. Eating for 8 hours or less may be feasible for some, or for many of us over a few days, but it becomes difficult for many people to sustain this over months or years. While the science at 12 hours is impressive, lowering your window (to as few as 8 hours) is significantly advantageous.

Time-restricted eating is never about counting calories; it is just about making you more disciplined about timing. We've found the best results for weight loss come with eating within an 8- or 9-hour window, and you can maintain this pattern until you get the desired results. Most of your body's fat burning happens 6 to 8 hours after finishing your last meal and increases almost exponentially after a full 12 hours of fasting, making any amount of time fasting past 12 hours highly beneficial for weight loss. Once you've achieved your desired weight loss, you can go back to an 11- or 12-hour window and maintain that body weight. Of course, discuss your plans with your doctor before beginning any new eating program.

A Typical TRE Day

Let's start by setting an ideal time for breakfast. The moment you eat breakfast or have your first cup of coffee or tea is the beginning of your eating window. Once you set your breakfast time, stick with it. If breakfast begins at 8:00 a.m., dinner must end by 8:00 p.m. We've found that it's healthiest to eat breakfast as early as possible. The reason is that the insulin response is better in the first half of the day and worse in late night. Besides, if you start early, you are also likely to end early, or at least 2 to 3 hours before going to bed. This is important, as melatonin levels begin to rise 2 to 4 hours before your typical sleep time. Finishing your meals before melatonin begins to rise is necessary to escape the interfering effect of melatonin on blood sugar.

The last few hours of nighttime fasting are very important. Imagine you're cleaning your house and you've put all the dirt in trash bags right by your front door. All of a sudden, a wind comes and topples over the trash bags and all your effort has been wasted. The same holds true if you eat earlier than usual in the morning. If your body isn't expecting a big flood of food coming in, all the effort that has gone on overnight to cleanse your system is for naught. This is particularly important when you are on a 12-hour eating cycle. If you are on a shorter eating interval of 8 to 10 hours, occasionally eating an earlier-than-usual breakfast may not blunt the benefit too much.

For the first 2 weeks, you can eat whenever you want within your window, but in between your first bite and your last it's a good idea to stick to a schedule of regular mealtimes. You are likely to find that as you adjust to eating for only 8 to 10 hours, when you wake up in the morning your metabolism and hunger will demand a bigger breakfast. Brushing your teeth in the morning (or at night for that matter) will not disrupt your TRE. Toothpaste doesn't count.

Breakfast is the meal that breaks your overnight fast. Don't be surprised if you are hungry in the morning. It's fine to eat slightly more at breakfast, especially if you choose healthy foods. Increase your fiber and protein intake in the morning. Having a full breakfast fills your stomach for hours. My go-to breakfast is a combination of oatmeal, cottage cheese, almond powder (that I make myself by crushing almonds in a coffee grinder), and dried cranberries. I travel often, and this is a very easy breakfast to take on the road.

An optimal breakfast is well-balanced and contains complex carbohydrates or fiber, lean proteins, and healthy fats. Fiber-rich foods are typically low-glycemic choices and will help you control your blood sugar throughout the day. Eating protein earlier in the day triggers the right amount of acid secretion in the stomach. So, instead of having more acid at night following a protein-rich dinner, you can switch to getting most of your protein in the morning and reduce your chance of getting heartburn and a poor night's sleep. This combination actually makes your digestive system work longer to digest food, and you will feel full (satiated and less hungry) for hours and will be less likely to snack on cookies, doughnuts, and other treats.

If your breakfast is substantial enough, then you'll probably feel full for roughly 4 to 6 hours. So if breakfast was at 8:00 a.m., you might feel a little bit hungry around 1:00 p.m. I've found that a salad or soup for lunch helps me get through the day. It provides great energy, and because it is light, I don't feel that postlunch lethargy that often follows a heavy meal. It will also hold me until I eat dinner with my family.

After breakfast, dinner is the second most important meal to align with your circadian rhythm, as it signals the end of your eating. Once your body recognizes that no more food is coming, it slowly transitions to repair and rejuvenation mode. You don't want to lose family time at the end of the day, and if you have a meaningful meal for dinner, then you have that time to spend together. Our study has also shown that people who follow TRE typically don't have that extreme hunger that they used to have at dinnertime. Over time, they are able to reduce this meal size.

In my home, we tend to have a traditional dinner of a protein and vegetables, cooked with healthy fats. We don't eat a lot of simple carbs at dinner because the body's glucose control is weaker later in the evening, and those carbs will be stored as body fat (you'll learn more about this in Chapter 10). After dinner, we make sure not to lie down or go to sleep right away. I give myself at least 3 to 4 hours between my last bite and going to bed for better digestion and better sleep.

You'll likely find that your system gets so used to the new timing within 2 to 4 weeks that you won't feel hungry after your target dinnertime. More surprising, people who have been on TRE for a while report that if they delay their dinner too late past their target time or have another drink or bite late at night, they can feel that the food just sits in their stomach, as if the stomach has been closed for the night and will return to work only in the morning. We like to refer to this as a *food hangover*.

Dinner Drinking

If you're going to have a cocktail, a beer, or a glass of wine, have it either before dinner or accompanying your meal. If you have an alcoholic beverage after dinner, it will postpone what is considered the last bite of the day: Even a sip is a bite!

Your body needs a lot of water throughout the day, especially if you work in a dry environment like an air-conditioned office building. Hydration has a circadian component. We are more likely to feel thirsty during the daytime because our body needs water to digest and process nutrients, make new building blocks for blood, and detoxify. It's a good idea to have a glass of water every hour or two so that you stay hydrated and energetic in the afternoon.

Drinking water after dinner doesn't disrupt your eating window. And if you wake up in the middle of the night and feel thirsty, go ahead and drink some water. I've found that if I don't drink water then I will stay awake, but if I have a drink of water it's likely that I will go back to sleep right away.

Lots of health books promote the benefits of drinking water, yet nearly 25 percent of people don't actually drink any water beyond what they get in their coffee. I don't count coffee as a water source as coffee itself can make us feel dehydrated and suppress sleep due to its caffeine content. But decaffeinated herbal teas may count toward water intake. Some people like a cup of herbal tea before bed, and as long as it doesn't have any caffeine, sweeteners, or milk, it's an acceptable option. Tea actually has a good amount of caffeine—the same ingredient in coffee that keeps us awake—and many herbal teas have caffeine. With a new brand of "herbal" tea appearing on the market almost every week, it is difficult to assess if it has caffeine or any other chemical that will keep you awake. So, I personally stay away from anything other than water after dinner.

Snacking Is Okay During the Day, but Not at Night

Daytime snacking is permitted, as long as you choose healthy options. The occasional birthday cake or cookie during the day is okay. In the evening, a small dessert can round out your dinner. But don't be surprised if your taste buds change once you start making better food choices. Slowly, you'll find that you are less enticed by overly sweet or salty foods.

After dinner has ended and the kitchen has been cleared, the eating

---

Late-Night Hunger and Stomach Cramping

Stomach cramps, especially the ones you might experience late at night, may stem from electrical activity in your gut gone awry. During the day, electrical activity (just like a muscle twitch) in the gut helps to move food through the intestinal tract. Research has shown that there is a circadian rhythm to this process, and it is now thought that people who have stomach cramps and indigestion actually have disrupted electrical activity. Instead of moving the food along in the right direction, if the electrical activity is even slightly altered, it might send the food in the wrong direction, causing pain or cramping.

In general, the gut's activity slows down at night. So, when you eat late into the evening, slow movement of food, or movement in the wrong direction, can cause stomach problems. This scenario is quite common. In fact, acid reflux medications are among the ten top-selling drugs in the United States, with more than 64 million prescriptions written in 2004 alone.[14]

Keep to Your Weekly Routine on the Weekends

The survey you filled out in Chapter 3 has already provided you with a glimpse of your current eating habits. We have found that most people don't realize that they eat for more than 12 hours a day. Others do a great job during the week, but their schedule falls apart on the weekend. This pattern is considered more than an "occasional" breach. For example, if you eat three times a week outside a 12-hour window, you are not adhering to TRE.

Remember, every time you eat, you turn on the entire digestive clock. As soon as food gets in your system, it must be digested, absorbed, sorted, and metabolized, and the waste must be sent to the kidneys and lower intestine. When you eat outside your window, even a small snack, almost all of your digestive organs must wake up from the rest and repair phase of the circadian cycle to digest and process the food. And once digestion begins, it will take several hours before the organs can go back to rest and repair mode. The next day when you start your morning bite at your usual breakfast time, your organs have to get back to work processing breakfast, even though they had an incomplete rest the night before.

This is exactly what happens when you change your eating window on different days. Your metabolic clock is automatically affected, as if you had traveled between two time zones within a week.

for the evening is done. You may feel hungry before bed, especially when you start eating within an 8- to 10-hour window. It's perfectly normal to experience these hunger pangs. You may even wake up from a deep sleep feeling hungry. Try hard to push past this by drinking a glass of water; as your body adjusts to this new rhythm, the late-night hunger will go away.

Eating late at night is by far the worst choice you can make, and it will totally defeat any benefits you've achieved throughout the day. First, snacking late at night disrupts the digestive clock: You reignite your metabolism in your gut, liver, and throughout your body. In this sense, you are literally waking the body when it is meant to be slowing down, cooling down, and getting ready for sleep. Although your brain told you that you're hungry, your organs are not ready to process the food.

A second problem is that because your gut was not prepared to digest the food, the food won't move as fast through your digestive system as it does during daytime. When food sits in your stomach, your stomach will secrete acid to digest the food. But if the food is not moving, this can cause acid reflux, especially if you try to lie down and go right to bed.

Swifty's Diet

Steve Swift first heard about my mice research in 2012 and contacted me then to see if I had done the same work with people. We were just beginning to think about a human experiment (ours wasn't done until 2015), so Steve decided to start one on, as he said, "the only body readily available to me: my own."

From that day onward Steve stuck rigidly to this diet. A little over a year later, Steve was back in touch with my lab. He had lost 72 pounds in 15 months. That was nearly one-third of his original weight! According to the BMI scale, he went from seriously obese to a normal amount of body fat.

Steve's TRE schedule is remarkable in its simplicity. He wakes up every morning at exactly 6:40 a.m. and eats breakfast at roughly 7:00. Eight hours later, he stops eating until the next day. He told me, "You can eat almost anything you like! I regularly have three puddings with my lunch. But aside from that, I do try to have some semblance of balance in my diet."

Steve reported to us that he has absolutely no side effects. He does get hungry before bedtime, but as he said, "never ravenously. I don't get cravings. But what I do get is about one hour of extra free time every day because I am not busy eating." I told Steve that I've heard similar feedback from others. Many people tell me that they enjoy having more time where they feel energetic in the evening, and they use that time productively with their families.

Steve is also experiencing other benefits. He told me that his knees, which had been painful for months, are not as achy and are giving him less trouble. I told him he could attribute that to the weight loss as well as a total body reduction in inflammation. He also told me that his memory is improving. Before TRE, Steve noticed that he was finding it harder to remember details like phone numbers, zip codes, dates, etc., and always had to write them down. Now, he no longer needs to carry around a notebook with these numbers.

Steve told me that with all this weight loss he was motivated to start running again. He's now up to 6½ miles per day, and he's using his bicycle more often instead of his car. Good habits beget more good habits.

FAQs

1. Is TRE for everyone?
Absolutely! The beauty of this program is that it forms the foundation for all health. Irrespective of region, culture, or cuisine, our ancestors ate all their food within a 10- to 12-hour window, and you can, too. And when you and your family follow this eating program together, you will all align to a single circadian code.

Children as young as 5 can be on a 12-hour TRE, and you'll see that it helps them stay in better health, sleep better, and avoid childhood obesity. Middle school and high school kids can also stay within a 12-hour TRE. Adults with hyperlipidemia, depression, high blood pressure, anxiety, or any other chronic disease may try a 12-hour TRE, but talk to your doctor before trying a longer fast.

Remember, TRE is not a diet. Diets are a protocol people follow for short periods of time to lose weight or address a health problem. TRE is a *lifestyle*. It is something you will want to do for the rest of your life. It is almost like brushing and flossing your teeth, where a simple routine takes care of most of your dental hygiene. However, you need to go to a dentist regularly for a better cleansing. Similarly, you may also want to try TRE for a shorter eating time, like a week of 8-hour eating every so often if you want to lose weight or improve your digestion.

Treat an 8-Hour TRE like a Holiday Meal

An 8-hour TRE reminds me of Thanksgiving. I can have a lovely meal early in the afternoon, and I'll feel full for the rest of the day. Have you ever tried to eat later after a Thanksgiving meal? When you do, that's when you go from feeling comfortably full to bloated from overeating.

---

2. Can I choose any 12 hours I want?

Any schedule you can follow is better than no schedule at all. However, there is an increased benefit to starting your window earlier in the day, as we've already discussed. We don't know for certain, but light may have some effect on our metabolism. For example, one study found that people who delayed their dinnertime did not see the same extent of weight loss as people who ate their dinner early.[15]

We do know that at night, melatonin level begins to rise to prepare our brain for sleep. Melatonin also seems to slow down our metabolism, and it acts on the pancreas, which produces insulin. This might be a mechanism to make sure the pancreas goes to sleep, because for millions of years, we didn't eat at night and it would be unnecessary to keep the pancreas up and running full steam throughout the night.

When your melatonin level begins to rise in the evening and you eat, the food triggers the insulin response to begin. The insulin helps your liver and muscles absorb glucose from your blood so that your blood glucose doesn't rise too high. But later at night, since insulin production is slowed down, there won't be enough to soak up all the glucose from the food. This will leave your blood glucose levels high for a long period of time. At the same time, your body might store the excess sugar in the blood as fat instead of using it as fuel.

3. Can I combine TRE with other diets?

Yes! If you've had good results following any diet—Paleo, Atkins, ketogenic, etc.—you can combine it with a smaller window for eating. TRE may in fact boost the benefits of some of these diets. For example, we have seen excellent results combining a strict 6- to 8-hour TRE with a high-protein ketogenic diet. Finally, the effects of caloric restriction can be boosted by adding time restriction and optimal timing.

4. Can I combine TRE and periodic fasting (5:2 diet) or a fasting-mimicking diet?

Any type of fasting 1 day a month helps you achieve an extra detoxification. You can easily combine TRE with a 5:2 diet, which consists of 5 days of regular eating and 2 days of restricted eating per week. On the days you eat, consolidate your meals to 12 hours or less. You may find that TRE is a great way to transition out of a 5:2 program once you've achieved your weight-loss goals.

5. Is there any downside? What are the potential dangers?

There may be some people who cannot tolerate 12 hours without food. I don't mean their stomach will grumble. Stomach grumbling with hunger is a sign that the stomach is empty and is ready for business. It also means that the body is switching from using readily available energy to tapping into its stored energy. But if you feel light-headed or dizzy after 12 hours of not eating, stop the program and talk to your physician.

Sometimes people take on a challenge too intensely, such as switching from eating within a 16-hour window to one that is only 8 hours. Or they eat very few calories and restrict their eating window at the same time. This combination can be very taxing on the body, especially if you are not used to a very low caloric intake. Instead, I recommend that you try to start your TRE with a 12-hour window without changing too much of what or how much you eat. After 2 or 3 weeks, try to reduce the eating interval or improve your diet.

6. What are the potential distractions?

We have found that there is a real 6-week hurdle to this program. This is the danger zone. After 6 weeks, you may start to see some changes to your weight, or not. If you don't see the results you're looking for, you might be disappointed or discouraged. However, this is the exact same time at which the hidden benefits begin. These benefits cannot be measured on a scale but may be found in better sleep, reduction in systemic inflammation, or improvement in motor coordination or overall energy level.

If you are following this program by yourself and don't get a real glimpse of these successes, it may be very tempting to stop. However, we know that we often adopt the lifestyle of our friends and others we hang out with. So, the trick is to start talking about TRE as soon as you see some benefits and try to convert the friends and family you hang out with or share meals with to adopt the program. Talking about your TRE benefits will make them aware of your eating habits, and if they notice these results, they will be more likely to try it themselves.

Most people adapt to a 12-hour TRE without much trouble. You can still share a breakfast or definitely share a dinner with your family or friends. If you want to do a TRE of 10 hours or less, eating with others may become a little tricky. However, if you can do a short TRE for a few weeks and then go back to an 11- or 12-hour TRE, you won't have to make drastic changes to your lifestyle for long. A shorter TRE is more beneficial for reducing body weight, reducing fat mass, and improving mood and endurance. Some people can maintain a TRE of 10 hours or less for months or years.

7. What about medications?

Medications are not considered food and should be taken per your doctor's orders. However, you can pay attention to what time you take your medication. Some medications actually work better when they're taken in the morning or at the end of the day. Talk to your doctor to see whether your current schedule is optimizing the outcome.

8. What about coffee?

Drinking coffee is one of the most difficult habits to get in alignment with your circadian code because it directly affects sleep. If you have a strong coffee habit, it might be a signal that your sleep is off. For instance, if you're addicted to an early morning cup or two to make sure you're fully awake, it is a sign that you're not getting enough sleep, or enough restful sleep, at night.

In our most recent research that follows the TRE patterns of shift-working firefighters and medical residents/nurses, we found that if these individuals stay awake all night or have fragmented sleep, their morning coffee is often used as a "safety drug" that helps them stay awake while driving back home. However, using a cup of coffee in this manner eventually backfires, because it prevents them from getting fully restorative sleep during the day. We suggested that they try to carpool or take public transportation so that they could get better sleep during the day and return in better shape to be productive at work the following day.

Even if you have your morning coffee by itself, without breakfast, it still counts as the moment when you break your overnight fast, so keep that in mind regarding your TRE window. Think about when you want to have that coffee, especially if you take it with cream and/or sugar.

Once you're addicted to coffee, you may need an additional caffeine boost in the late afternoon. This second round is very likely to interfere with sleep. Coffee can stay in your system as long as 10 hours. That's why the conventional wisdom is to avoid coffee past noon. If you are having a lag in energy in the afternoon, it's possible that you're dehydrated: Try a glass of water and see how you feel.

9. Can I stay on TRE forever?

Absolutely! You may not want to stay on an 8-hour TRE forever, but you can easily stay on a 10- to 12-hour TRE as a lifestyle. Your circadian code will stay robust and your chance of getting a chronic disease will stay low!

10. How often can I cheat?

We really don't think of TRE as something you can cheat on. However, when you get off track, get right back on. You can still reap the benefits of TRE with an "off day" once in a while. While off days will disrupt your circadian code, being on TRE for 5 or 6 days in a week is much better than randomly eating all week.

Let's say on Monday through Friday you did great, but on Saturday night you went out with friends and blew your whole schedule. Don't panic! If your last bite of food (or drink) on Saturday night was at 11:00 p.m., you can still get back on track the next day. In fact, it's quite likely that you won't feel like eating breakfast at your usual time. Listen to your body. If you're not feeling hungry, don't eat. When you finally feel hungry, go ahead and have your first meal. If that first meal is close to noon, consider it lunch. Have a well-balanced meal and then try to get back on track with your dinner. If your target is to have your dinner by 7:00 p.m., do that and go back to your original plan.

Next time, consider happy hour. The food is cheap, and you won't blow your TRE!

Chart Your Progress

You can use the following chart to track your TRE. For one month, write down when you took your first and last bite each day, and then the next morning, record how many hours you slept that night. First, notice if your sleep is improving, and how that improvement is correlated to your TRE. Are you getting the best night's sleep with the tightest TRE? Or is just consolidating your eating within 12 hours doing the trick?

Then, track how the rest of your health is changing. It may take a week or so before you notice if your health, mood, or energy improves. You might also see that you hit a plateau, then overcome it later in the month. This is a very typical pattern we've found in our study.

Copy this chart to use again and again, or simply transfer the data into a normal calendar. Research shows that keeping an accurate record of your health is one of the best ways to stay on track (as you learned in Chapter 3). Or, you can use the myCircadianClock app by signing up at mycircadianclock.org.

Month one	First bite	Last bite	Hours slept	Noticeable changes in health, mood, or energy
Day 1				
2				
3				
4				
5				
6				
7				
8				
9				
10				
11				
12				
13				
14				
15				
16				
17				
18				
19				
20				
21				
22				
23				
24				
25				
26				
27				
28				
29				
30				
31				

Christine Couldn't Sleep

Christine had a lifelong struggle with sleep. Ever since she was a child, she didn't remember ever getting 7 hours of sleep in a night. She tried everything in our sleep protocol, including managing light, keeping a dark room, and doing exercise during the day, but nothing worked. She tried different medications and was taking sleeping pills to go to sleep, but they made her feel groggy all day. After 6 years of taking sleep medications, she wanted to give TRE a try.

We gave Christine an activity watch to measure her activity and sleep, told her to ditch the sleep meds, and had her start an 8-hour TRE protocol. The first week she reported that she was hungry at night and could not sleep. Most people would have given up at this point, but Christine was desperate. Finally, by day 8, she found that she was actually sleeping better. By the end of the second week, for the first time in many years, she could sleep 5 to 6 hours without meds. She admits it is hard to stay away from food after 6:00 p.m., particularly when she has to meet with friends or family, and she does go outside this window once in a while to enjoy a small snack with friends. But she is confident that she has one more tool other than pills to get a good night's sleep.

What to Eat

There's no getting around the fact that TRE requires a bit of planning. You won't be eating around the clock, so you may want to plan your meals carefully so that you don't ever feel ravenously hungry. At the same time, I can't predict what 12-hour window you should start with. Some people love breakfast foods and need them to jump-start their day. Others wait until noon and can more easily handle a shorter TRE. Only you can make that decision. In the next chapter, you'll learn that your brain really doesn't need breakfast to provide "extra energy." It's totally up to you.

For the best weight-loss and overall health results, follow a balanced diet: plenty of fresh fruits and vegetables, lean proteins, and healthy fats.

Remember, you're not counting calories. At the same time, don't go crazy on doughnuts or deep-fried anything. Here's a great list of seven foods you really should stay away from. Think of these as the 7 Rules of Successful TRE People.

1. No soda, diet or otherwise. Drinking full-calorie soda is one of the easiest and most effective methods for streaming sugar into the body and disrupting your blood glucose system. Drinking soda is one of the most obvious ways people overconsume calories. And diet soda is not a healthier alternative. It is thought to change the gut's microbiome (which you'll learn more about in Chapter 9),[16] and you need all the good bacteria you can get.

2. No prepackaged fruit juices or vegetable juices. Even the ones that say they are "100 percent fruit juice" are not great choices because most of them contain preservatives that can corrode your intestinal lining, causing leaky gut syndrome (which you'll learn more about in Chapter 9). If you must have fruit or veggie juices, make them yourself. And drink them the same day.

3. No breakfast cereals—unless they have fewer than 5 grams of sugar per serving. You don't need to start or end your day with sugar.

4. No "energy," protein, or any variety of fruit-and-nut bars. They're just candy bars, even if they are marketed with triathletes or sports stars. Many of them have tons of protein and fiber, but most have lots of preservatives and sugar. You're always better off eating a handful of nuts instead of having them pressed into a bar.

5. No processed foods that contain corn syrup, fructose, or sucrose (50 percent of sucrose is fructose). Read labels carefully as these ingredients can be found in anything from pasta sauces to candy bars. You want to avoid them because even though they are used as sweeteners, the body doesn't recognize them as forms of sugar, and they trick your blood glucose control system into reacting as if there is no sugar in your blood, causing your blood sugar levels to rise. This is a big problem for everyone, especially if you are already diagnosed with prediabetes or diabetes.

6. No dark chocolate/hot chocolate in the evening. One 5-ounce bar of dark chocolate has the same amount of caffeine as a cup of coffee. If you love your chocolate, have milk chocolate, which has half the caffeine as dark chocolate, and eat it right after lunch.

7. No commercially processed nut butters. I love peanut butter as much as anyone. Look for kinds that have one ingredient: the nuts. Skip anything with added sugar or oil.

Vegetarians Need to Choose Proteins Carefully

Vegetarians often eat lentils as a source of protein. However, lentils have around 25 percent protein and are almost 65 percent complex carbohydrates. So, while they're a healthy choice and will keep you feeling full, they're not a high-protein choice. A better vegetarian protein option is tofu or cottage cheese.

The Importance of Proteins

Foods that are high in protein contain vital amino acids, which are the building blocks necessary to produce various enzymes and create muscle. All plants and animals require amino acids, which is why they occur in all food sources. Plants can create amino acids through sunlight and water, while animals (including humans) can create only some. There are additional amino acids we must get from the foods we eat.

You can enjoy all different types of high-quality protein sources. The foods highest in protein include animal meats, poultry, fish, seafood, beans and peas, eggs, soy, nuts, and seeds (you'll find a complete list on pages 118–119). Leafy green vegetables as well as dairy also contain protein. Animal proteins are the richest source of protein.

Is it possible to eat too much protein? Yes. The rule of thumb is your daily intake should be 0.36 grams of protein per day per pound of body weight. So, for someone who weighs around 150 pounds, that would be about 2 ounces of protein a day. Look carefully at this recommendation: Most of us are eating plenty of protein. However, too much protein (more than 1 gram per pound of body weight for several weeks or months) is not good for your health. Excessive protein intake stresses your metabolism, which is hard on your kidneys, and you really would like to go through life with two working kidneys.

Protein drinks sound like a good idea to build or maintain muscle mass, especially if you want additional support for exercise. However, they can also contain a lot of ingredients you don't really want to eat. For example, one shake mix might have 15 grams of protein and 10 grams of sugar to make it drinkable. If you feel you need a protein drink, choose one that doesn't contain added sugar.

Choose Complex Carbohydrates

The healthiest carbohydrates are found in nonstarchy, leafy green vegetables and fruits and grains that are low on the glycemic index (GI). This is a rating system that evaluates how different foods affect blood sugar levels. High-GI carbohydrates spike your blood sugar, causing a flood of insulin that triggers your body to store fat and become hungry again within several hours. In contrast, slow-burning, lower-GI foods, such as oatmeal and green vegetables, cut your appetite. These lower-GI carbs are more efficient at keeping glucose levels stabilized and insulin in check. Low-GI fruits include berries and citrus fruits.

Limit or eliminate simple and processed carbohydrates, such as white breads, pasta, white rice, pastries, cookies, and cakes. Instead, choose whole wheat versions that are also high in fiber. Foods high in fiber are predominantly carbohydrates, but they are good choices because your body can't digest fiber and it scrubs your intestines as it leaves the system. Fiber helps detox your body and provides nutrients for a healthy gut. Beans, berries, leafy green vegetables, quinoa, and whole grains are all good sources of fiber.

> ### Best Choices for Rice
>
> In my family, we have switched from traditional basmati rice, which is highly processed and has a high GI, to parboiled rice. This is considered a complex carbohydrate because it is difficult to digest. It has the same healthy components as brown rice, which is another good alternative to white rice.

Good Sources of Healthy Fats

Dietary fat provides your body with the building blocks for every cell. You need dietary fat for brain development and to keep your skin and hair healthy. Fat also helps you absorb important micronutrients, including vitamins A, D, E, and K. Lastly, adding fat to meals helps us feel satisfied and keeps us feeling full longer.

The healthiest fats are those found in whole foods, as opposed to fats from processed oils. Saturated fats, like butter, remain solid at room temperature. Saturated fat is not unhealthy or fattening, despite what you might have been told. Monounsaturated fatty acids are the best fats and are liquid or soft at room temperature. They are found in foods like olive oil, avocado, nuts, seeds, and egg yolks. Monounsaturated fats are a featured part of a Mediterranean diet, which is thought to keep people healthy and slim. Monounsaturated fats are also easy for our body to use as energy.

Polyunsaturated fatty acids, found in many plant and animal foods, are also a good source of fat. There are two types of polyunsaturated fats: omega-3 fats and omega-6 fats. Omega-3 fats play a pivotal role in maintaining good health and can also help control and reduce body fat. This is because omega-3 fats have the ability to increase blood flow so fats are more easily delivered to the sites where metabolism is stimulated.

Omega-3 fats are found naturally in a few plant foods, such as flax; they are also found in a few fish, such as salmon, in shrimp, and in some eggs.

Omega-6 fats are found in the highest quantities in vegetable oils, such as corn, soybean, and safflower oil. Omega-6s also make up all the polyunsaturated fat found in chicken, beef, and pork. Because omega-6s are found in so many of the foods we normally eat, we usually get enough of these fats to meet our dietary needs. Both types of polyunsaturated fats are referred to as "essential" because your body cannot make them itself, or work without them.

The Circadian Code Shopping List

Low-Glycemic Fruits and Vegetables

Apples	Cauliflower
Apricots	Celery
Artichokes	Coconuts
Arugula	Collard greens
Asparagus	Cucumbers
Avocados	Eggplant
Bananas	Fennel
Beet greens	Fiddleheads
Bell peppers	Figs
Blackberries	Garlic
Blueberries	Grapefruit
Bok choy	Jerusalem artichokes
Broccoli	Jicama
Brussels sprouts	Kale
Cabbage	Kiwis
Carrots	Leeks

Melons	Raspberries
Mushrooms	Romaine lettuce
Mustard greens	Rutabaga
Olives	Sea vegetables
Onions	Spinach
Parsnips	Squash
Peaches	Strawberries
Pears	Swiss chard
Peppers	Tomatoes
Prunes	Turnip greens
Pumpkin	Watercress
Radishes	

Protein from Animal Sources

Beef	Lamb
Bison/Buffalo	Pork
Chicken	Turkey
Duck	Veal
Eggs	

Vegetarian Proteins

Black beans

Black-eyed peas

Garbanzo beans
(chickpeas)

Kidney beans

Legumes

Lentils

Navy beans

Peanuts

Pinto beans

Split peas

Sugar snap peas

White beans

Fish and Shellfish

Catfish

Clams

Cod

Crab

Crayfish

Flounder

Haddock

Halibut

Herring

Lobster

Mackerel

Mussels

Octopus

Oysters

Pollock

Salmon

Scallops

Sea bass

Shrimp

Snapper

Squid (calamari)

Swordfish

Trout

Tuna

Nuts

Almonds	Pecans
Brazil nuts	Pine nuts
Chestnuts	Pistachios
Hazelnuts	Walnuts
Macadamia nuts	Derivative nut butters

Seeds

Chia seeds	Pumpkin seeds
Flaxseeds	Sesame seeds
Hemp seeds	Sunflower seeds

Healthy Fats and Oils

Avocado oil	Macadamia oil
Butter	Olive oil
Coconut oil	

CHAPTER 6

Optimizing Learning and Working

Everything you do, all day long, requires learning; this remains true throughout your life. Children learn at school, and adults learn new life skills or improve their performance on the job. At home, we are constantly learning how to be better parents, partners, friends, coaches, and even cooks.

Every task we master through learning involves both the brain and the body. In fact, I believe that there are seven criteria for learning. Each is influenced by our circadian code and is affected by either an optimal exposure to light, an optimal amount of sleep, an optimal eating schedule, or a combination of these factors.

Attention

Attention is the ability to stay focused and complete a task without distraction. Attention also requires adaptability, the ability to withdraw from one activity in order to deal effectively with another. Children at school have to pay attention to what is being taught so that they can imprint it in their working memory and then consolidate and transfer information into long-term memory storage. The same is true for adults: Unless we pay attention, we cannot create memories. For example, if you are a banker or stockbroker, you pay attention to how stock prices move, integrate that information into your working memory, decide what to do, take action, and remember the event so that you can perform better in the future. The same thing is true for a physician, pilot, air traffic controller, truck driver, artist, homemaker, etc. Attention also requires a precise amount of concentration: too much and you will not be able to get off a task; too little and you will not be able to start one, and you certainly won't be able to finish.

Attention has a circadian component. We have an internal drive to be more attentive during the day and are naturally prone to being less attentive at night. However, sleep deprivation messes with your attention. A sleep-deprived brain cannot stay focused on tasks during the day, as the biggest distraction is feeling sleepy and dozing off.[1]

Working Memory

Working memory is the most important function of the human brain; it separates us from all other animals. It involves the ability to absorb information, retain it, and connect it to information you have already learned. For example, as you drive down a street, you are using the right amount of pressure on the gas pedal and at the same time you are observing the cars in front of you and the landmarks you're passing, all the while trying to figure out where you are going. When your working memory is functioning at a high level, you perform well at home and at school. When it is low, you feel scattered, forgetful, and sometimes anxious.

Sleep deprivation compromises your working memory by affecting your reaction time. When you see something new, you observe that information and use your memory before you take an action. For example, if you are driving on the highway and the car in front of you stops, poor sleep will compromise your reaction time and can lead to an accident. We know that most car accidents happen in the morning. We also know that many large-scale accidents, such as the *Exxon Valdez* oil spill and the Chernobyl nuclear power plant explosion, are related to sleep deprivation.

Positive Reward Assessment/Negative Reward Assessment

Positive and negative reward assessments are how we use attention and working memory to make decisions. For example, you've already learned and stored in your memory that fresh fruits and vegetables are healthy snacks (positive reward). You also know that potato chips are a bad choice (negative reward). But you go to the grocery store and there is a huge sale on chips. And you like chips. If you have slept well, and you're hungry, you're more likely to make the positive reward choice and buy yourself an apple or a banana. But if you're sleep-deprived and hungry, you are just as likely to buy the chips even though you know they aren't a healthy choice.[2]

Positive and negative reward assessments also influence how we communicate. When we communicate with anyone, we have a good idea about what will make them happy and what will make them upset. Without good sleep, we're likely to say something we'll regret. In that way, sleep deprivation affects our relationships.

Hippocampal Memory

The hippocampus is part of the brain's most primitive area—the limbic system—and it plays an important role in the consolidation of information from short-term memory to long-term memory. Hippocampal memory involves calling up information you learned last week and applying it to the task at hand. One of the major functions of sleep is memory consolidation in the hippocampus.[3] For instance, if you are learning a new language, starting a new math chapter, or playing a new video game, you are more likely to master the skill if you've had sufficient sleep than if you've had a few sleepless nights.

Long-term memory suffers with sleep deprivation. At first you may feel like you are more forgetful, but over time, you'll have a harder time storing even new memories, which affects recall necessary for learning and working.

Alertness

Your brain is most alert in the morning. As the day goes on, the circadian clock instructs your brain to be less alert: This is why some people complain of losing focus at work toward the end of the day. Around 9:00 or 10:00 p.m., the alertness drive actually reverses: There is minimal drive to stay alert, and we go to sleep. That's when your brain switches from active control to a default mode. It no longer has to listen to your commands and instead goes on autopilot for repair, strengthening its neuronal connections, and moving memories from working memory to hippocampal consolidation.

Mood

Mood is our state of mind—whether we are feeling happy, energetic, low, anxious, irritable, angry, etc. Our moods can be transient and can change based on what we experience in everyday life. It is normal for happy news to boost our mood, while a sad event might make us feel low.

Having less sleep disrupts a normal response to events and makes us susceptible to more extreme mood swings; we tend to be more irritable, anxious, and angry. What's more, for most people, sleep deprivation tilts their mood toward a negative state.

One of the natural factors that most influences mood is light. Have you ever noticed that if you spend a day in a dark room, your mood may be low and your mind foggy, even if you have had a good night's sleep and ate well the day before. You might not feel quite like yourself until later in the morning, when bright natural light enhances your mood. An animal study done at Johns Hopkins University showed insufficient light triggers depression-like mood and impairs learning in mice, and this effect was connected to insufficient activation of the blue light sensor, melanopsin.[4] Similarly, a unique collaborative study between neurologists and architects found that office workers with access to daylight had better mood, performance, and sleep quality than those working in windowless offices.[5]

Autonomic Function

The brain is composed of a *central nervous system,* where all active learning takes place; a *peripheral nervous system,* which connects the brain to organs, including muscles, in order to control movement; and an *autonomic nervous system,* which controls everything that happens automatically, like breathing, heart rate, digestion, and the production of stress hormones. For learning and working to occur optimally, we need all three domains to be in top form, including the autonomic nervous system. If your heart rate is not right, you may experience palpitations; if your digestion is off, you may get a stomachache; and if your stress hormones are too high, you may feel stressed out. These conditions are distracting at best and anxiety producing at worst.

Each autonomic nervous system function has a circadian component. At nighttime, autonomic activity subsides, so heart rate, breathing, stomach movement, and even stress hormone production slow down so we can go to sleep. During the day, autonomic activity is at its peak, and so is our ability to work and learn. However, chronic sleep deprivation or an interrupted sleep routine can increase the level of stress hormones produced or sensitize our stress system such that we automatically overreact in response to minor stressors.[6]

Many of these same hormones, as well as bacteria normally found in the gut, can affect brain function and mood, causing panic attacks or anxiety if the master circadian clock is broken.[7,8] As we will learn in Chapter 9, time-restricted eating strengthens the daily rhythms of the gut and can restore a normal balance of gut hormones and bacteria, improving brain function. TRE also improves the daily rhythm in the brain's autonomic functions so that the right amount of stress hormones are produced, which then improves mood.

The Optimal Workday

When these seven factors are at their peak, your ability to get work and learning done is high. Good learning and performance is typically a sign that you are aligned with your circadian clock, and yet there is always room for improvement. The next place to investigate is how well you are aligned with your circadian code.

Your optimal brain function is highest between 10:00 a.m. and 3:00 p.m.; that's when you should notice that your best work or learning is done. Studies have shown that this is the window during which we are in the right frame of mind for making good decisions, solving multifaceted problems, and navigating complicated social situations.

The rising phase of peak performance starts at 10:00 a.m. and tops off around noon. It is during these few hours that your brain is really performing at its peak: Your attention, working memory, assessments, and mood are at their highest levels. From noon onward, your brain begins to slow down. This is a good reason not to lose an hour of top productivity by taking a long lunch. In fact, long lunches work in opposition to your circadian rhythm. If you work through lunchtime, or take a brief lunch break, I have found that productivity increases such that the same number of tasks that used to take 8 hours can be finished in 7.

Toward the end of the day, the brain gets tired and we cannot do complex, complicated work as well as we could earlier in the day. This is further worsened by two factors that most people experience. As we discussed earlier, insufficient sleep from the prior night increases the sleep pressure as the next day progresses, so if you had less sleep the night before, by afternoon your brain is feeling that extra sleep pressure. In addition, if you had a heavy lunch, you are more likely to feel sleepy for up to 2 hours afterward.[9] If your typical lunchtime is between noon and 1:00 p.m., you'll notice that your attention and mood start to taper down around 3:00 p.m. Yet if you've optimized your morning and early afternoon hours, you will have already gotten your work done.

If you had insufficient sleep and a heavy lunch, you may try to fight the afternoon slump with a snack. However, as we discussed earlier in terms of positive and negative rewards, a sleepy brain is likely to make poor food decisions. The problem is that unhealthy sugary treats will boost your energy for only a very short amount of time and will ease your hunger only in the short term. You may find that you need another treat later in the day to stave off hunger until dinnertime. So, while it seems like an effective strategy in the very short run, it's not a good one.

Instead of going for a sugary treat to wake you up later in the afternoon, try a glass of water or some hot decaffeinated tea, a piece of fruit, or a handful of nuts. The glass of water is your best bet, though, because there is a circadian rhythm for hydration, and our body requires that we drink water during the day, even though many people neglect this urge. If you're feeling tired in the afternoon, your body may be trying to tell you that that you are dehydrated.[10] If you have a drink of water, you'll be surprised how much more energetic you'll feel, without adding more, very empty calories. If you can build this into a habit, you'll never reach for that 3:00 p.m. doughnut again.

Working in a windowless office or doing monotonous tasks can also cause fatigue. Break up your day by taking a short walk outdoors; it might perk you up so that you can get through the rest of the day. Even standing and stretching every hour can help you stay focused.

Sometimes people want to go back to work after dinner or power through the day and stay at work as late as possible. You know those folks, or maybe you are one of them: They equate staying longer in the office with being a better employee. However, two things are happening to your circadian rhythm that actually make you less productive during the evening hours. The first is that your natural sleep drive is increasing and your alertness drive is decreasing. Second, you're likely working in a darker room than during the daytime, and the dim light has a different effect on your brain: It literally makes you foggy, so your brain cannot think clearly. No matter how hard you try, you simply can't force your brain to learn and work optimally at night. You may crank through a few nights, but it is not sustainable.

Now, you may be thinking, *This is great to know, Dr. Panda, but my child has 5 hours of homework every night* or *I'm a shift worker* or *I'm constantly pushing against tight deadlines at work.* How can we hack the circadian code to increase our productivity?

Let's explore the three key components of sleep, light, and timing to see what you can do to optimize your circadian code and be more productive, given the realities of your life. My three biggest tips are:

- You have to give up the notion that staying up more hours will make you more productive. In fact, the opposite is true. If you set aside 8 hours for sleep opportunity (total time including sleep and preparation for sleep) to prepare for a productive day, you are giving your brain the rest it needs to be ready for the next day.

- During the day, optimize your productivity with natural light exposure to keep you more alert and productive.

- In the evening, adjust your light exposure to prepare your brain for restorative sleep.

Master Light, Master Productivity

For the vast part of human history, our ancestors spent most of the day outdoors with plenty of exposure to natural daylight. Even if they were under the shade of a tree or a cloud, they still received plenty of bright light, in the order of thousands of lux. A lux is a unit of measurement that signifies the amount of light that is received by the eye. Daytime outdoor light is typically measured between 1,000 lux (a cloudy day) and 200,000 lux (full sun in a desert). An office without windows is typically between 80 and 100 lux; a home using overhead lights can be as low as 50 lux. The figure on page 129 will give you a fair estimate of the amount of light found in different types of buildings and how it relates to our circadian rhythm and mood.

In modern times, an average person spends more than 87 percent of their time indoors; we average only 2½ hours outdoors, half of which is often after sunset. Our indoor light environment may be disrupting our circadian rhythm and compromising our mood. Yet we know that when it comes to enhancing learning, memory, and working, we have to pay attention to light. Our circadian rhythms are designed to adapt to the natural cycle of light and darkness. Your brain needs light to turn on all of its functions.

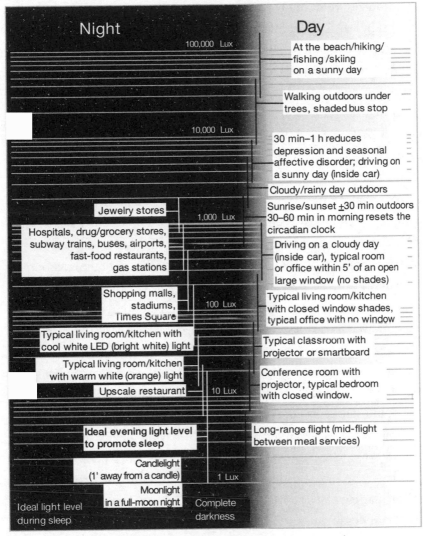

The different amounts of light we are exposed
to in different environments.

When you wake up, bright light is detected through the eye's blue
light sensor, melanopsin, and when that happens, melanopsin tells the
brain to stop producing the sleep hormone melatonin and start increasing production of the stress hormone cortisol, which will help you begin
to feel alert and ready to start the day. Bright light in the morning also

synchronizes your brain clock to the daytime so that your circadian rhythm in learning and memory will begin to rise and you'll reach optimal productivity a few hours later.

As I mentioned, we know that increased exposure to light leads to improved mood. We also know that improved mood leads to increased performance. Therefore, is it possible that increased light leads to increased performance? The research points to this as fact. Bright lights used during the day in the office or at home have been found to improve mood, alertness, and productivity.[11,12]

No matter where you live, we know that if you limit your exposure to natural daylight, you're more likely to experience low moods and have trouble making good decisions. The reason is that significant daily exposure to artificial light disrupts the circadian code: Your office or home environment will rarely provide enough lux to match even the cloudiest daylight. However, your work or learning environment can be optimized by simulating daylight. And if you are exposed to some natural light early in the morning, all the better. You need at least 1 hour of daylight exposure—being outside, driving in your car, sitting by a window where you can soak up *at least* 1,000 lux of light—to reduce sleepiness, synchronize your clock, perk up your mood, and stay happy and productive throughout the day.

One way to access more daylight is by having breakfast by a window, or if the weather permits, having breakfast outside. Walking to work or school also increases daylight exposure. Parents can even drop their children a few blocks away from school so that they can get at least 15 to 20 minutes of direct exposure to outdoor daylight before school starts. Small changes lead to big results.

While it's ideal to have access to daylight in the morning, having some outdoor time any time during the day is better than none. If you or your child can eat lunch outdoors, or if your cafeteria or your kitchen has large windows that let in a lot of light, it is better than nothing. However, we cannot accumulate and store light exposure to be used at a different, later date. Daylight needs to be available during the daytime, when we actually need the light to keep us alert and to learn.

Whenever you are indoors, choose to sit right next to the largest window available. You might get 2,000 to 5,000 lux of light on a good day, but if you move 6 feet away from the window, the light may only be measured at 500 lux: a significant difference. And if your windows are covered by blinds or screens, your daytime indoor lighting may be 100 lux or less. Even the best, brightest LED light bulb only throws 1,000 lux.

The bottom line is we want to boost light when we are awake (typically during the day) and reduce light—specifically from the blue spectrum—during the night (or for at least 8 to 9 hours around our sleep). Unlike a few decades ago when all sources of light were primarily from light bulbs at home, we now get significant exposure from digital displays. Therefore, managing light for circadian rhythm involves managing light sources, including digital devices. In fact, when you are working on a computer or tablet, the amount of light that is emitted from a typical screen over 1 to 2 hours is sufficient to suppress your evening melatonin and disturb sleep.[13,14] However, there are new technologies that automatically reduce brightness or color of computer screens and smartphones at a set time. You can use these settings to reduce exposure to circadian-disrupting light from digital screens in the evening.

If you have to work later into the evening, hack your light. You'll be most productive if you can switch to task lighting that lights up only

Living in Beijing Time

China operates under one time zone: Beijing time. So, at 8:00 a.m. in Beijing, which is situated in far eastern China, it is bright and sunny, while the western extreme of China is still very dark. Adults in western China who have government jobs or do business with the eastern part of the country have to wake up in total darkness so that they can work on Beijing time. At the same time, they try to have a normal family life, which makes it difficult to go to sleep before 9:00 p.m. This lack of time zones breaks their circadian rhythm.

your work area and reduces direct exposure to your eyes, as compared with overhead or horizontal lighting.

But more important, don't let your work disrupt your sleep schedule. There's no way you can be productive and tired at the same time.

The Truth about Food and Productivity

Eating at the same time every day is one of the most powerful ways to maintain a strong circadian rhythm. This is true for breakfast and dinner especially. Between these two meals, it's less important to focus on when you eat and more important to focus on eating foods that support a healthy brain. When it comes to brain function, quality is more important than quantity. Eating more food doesn't mean our brain will function better. The brain actually works better on an empty stomach. We are not very alert after we finish a meal. This may relate to our inherent survival strategy. When we are hungry, our brain must find creative ways to search for food.

Your performance at any moment during the day is primarily determined by what you did the night before—when you ate and how much you slept—because that is what sets your clock, which then primes your body and brain. Studies have shown that both modest fasting and exercise have a similar brain-boosting effect. Each of them can increase a chemical called *brain-derived neurotropic factor* (BDNF) that improves the connection between brain cells and improves brain function.[15,16] When you have plenty of BDNF combined with a good night's sleep, your brain is better prepared for performing complex tasks, staying focused, and being productive, so you can complete the same amount of work in less time.

Eating a late-night meal negatively affects your ability to pay attention the next day. As you learned in Chapter 5, late-night meals or middle-of-the-night snacks disturb our circadian clock, and as a result, the peak performance window between 10:00 a.m. and 3:00 p.m. is disturbed.

The Coffee Productivity Myth

The active ingredient in coffee is caffeine, which does not have any nutritional benefit; our body doesn't need caffeine to function. Caffeine is found naturally in more than four dozen plants and pods, including the coffee bean, tea leaf, kola nut, and cacao bean. People consume caffeine in many forms—coffee, tea, cocoa, chocolate, soft drinks, energy drinks, and some over-the-counter medications. Typically, 100 to 200 milligrams of caffeine (found in three 8-ounce cups of medium-roast coffee, or two to three bars of dark chocolate) is considered moderate daily intake. One cup of tea can have 25 to 30 milligrams of caffeine.

Caffeine is a stimulant, and in a low to moderate dose, it can increase alertness and reduce sleepiness. For the average person, the effect is almost instantaneous: Most of the caffeine is absorbed within 15 minutes and can start its stimulating effect in that time frame.

Although coffee can improve alertness, it does not remove your sleep debt. Rather, it delays that sleep pressure to a later time. That is why sleep-deprived people tend to have a "caffeine crash" after the effect wears off. They need another dose of coffee to keep them awake. In the evening, caffeine intake combined with light exposure can further delay your sleep. Even though a study published in the *British Medical Journal* declared in its headline that "Coffee Gets a Clean Bill of Health,"[17,18] some caution is urgently needed. The article also states that the health benefit of coffee is correlative and no formal cause is established. It also cautions that physiological effects of coffee on increased heart rate, stimulation of the central nervous system, and feelings of anxiety were not taken into consideration. Other pro-caffeine reviews also excluded studies that focused on the adverse effect of coffee on sleep quality or duration[19] and did not mention other studies that have shown coffee can compromise how our body handles glucose[20] or how coffee directly disrupts our circadian rhythms.[21] An increasing problem in the United States is that popular "coffee" drinks have become sizeable 16- to 24-ounce concoctions of syrups, whipped cream, milk, and caramel sauce topped with coffee.[22] Coffee in this form is a vessel for empty calories in the form of added sugar. Overall, coffee does work as a quick fix to hack your sleepiness, but it's not an ideal choice for optimal health.

If you find that you feel sleepy after a big lunch and need another cup of coffee to stay awake, you might want to shift your bigger daytime meal to the morning and have a lighter lunch that will make you feel less tired. In the morning, your alertness drive is at its highest point and your circadian sleep drive is minimal, so you are less likely to be affected by a heavy meal in the morning. And without the extra cup of coffee after lunch, your sleep pattern won't be disrupted.

In the evening, if you and your family finish dinner together before 6:00 or 7:00 p.m., you have the entire postdinner evening to properly digest your meal. As long as you reduce your light exposure, you will slowly build up your sleep drive and will find it easy to fall asleep without using a sleeping pill or a nightcap cocktail.

Lack of Sleep Disrupts the Circadian Code for Learning

Sleep loss leads to four major effects on our circadian code. First, lack of sufficient sleep doesn't give our brain enough time to consolidate memories. Second, staying up late reduces brain function and productivity that night. Third, when we sleep less, we are exposed to additional light and the opportunity for eating in the middle of the night, both of which disrupt our circadian clock. Fourth, the following morning, as we wake up late and rush to work, we have little time to get the right amount of morning light exposure that will brighten our mood.

Having less sleep also has a direct impact on the maintenance of our brain network. One study conducted with well-controlled conditions[23] showed that if you take a person who sleeps for 8 hours and give them the same math lesson every day for a week, by the end of the week this person will have mastered that particular lesson, going from scoring 10 on a scale of 1 to 100 to scoring 100. But if the person sleeps only 4 hours, they go from 10 to 50. They master only half the material.

Outside the lab, we see the same effect in real life. Ben Smarr studied almost 300 students in Seattle.[24] Most of these students were taking the same college biology class. He gave them access to an online sleep log on his website, in which they kept track of what time they went to bed and what time they woke up over the course of a month. He then analyzed how their sleeping patterns affected their grades. As you can imagine, there was a relationship between having a good night's sleep and getting better grades. Specifically, not adhering to a regular bedtime was correlated with lowered academic performance in both men and women, and women were found to be even more sensitive than men to changes in sleeping patterns.

School Start Times and the Circadian Code

School start times are becoming a hot topic across the United States, and I believe our kids deserve every minute of extra sleep they can get. A lot of scientific evidence supports the idea that high schools should start later in the morning.[25,26,27] Delayed start times would positively affect students' circadian code and would improve their alignment of all three factors: light, sleep, and food.

As we discussed earlier, adolescents are most sensitive to evening light, which delays their clock and bedtime. Their circadian clock does not wake them up early, yet schools are all scheduled to start early, sometimes before sunrise. This produces a conflict between the circadian clock and what the education system demands. An early start time also causes students to miss morning light, which throws off their natural circadian code.

When teens are sleep-deprived, they make poor food choices. When they are rushing out the door in the morning, they might grab a cereal bar instead of a proper breakfast. These bars tend to be laden with sugar and will not support a full day of learning.

Extracurricular activities, including sports practices, in the evening under superbright lights can also affect your child's circadian code as well. These bright lights can suppress their natural melatonin production, delaying their circadian clock and keeping them awake late into the night. It's no wonder that these kids are not going to bed before midnight. So, it's not only the school start time. The after-school activities are another factor that affects our children's circadian code.

The Problem with Smartboards

In the past 10 years, the classroom has changed. There are fewer white-boards and blackboards, as schools have invested in smartboards and pro-jectors with large display screens. Teachers then keep the rooms dark in order to use the overhead projectors. This is a dangerous trend because it further limits daylight exposure.

Brighter Office Buildings Yield Better Results

My lab was contacted by an architecture firm interested in improving mood and productivity through healthy building design. They had heard about our work on melanopsin and its relationship to sleep, mood, and alertness. They quickly realized that their own office building was very dark and that very few employees had access to their own windows. As they began to research new buildings, we showed them how to choose one that offered more access to daylight. They also wanted to measure if access to more daylight would improve mood and sleep at night.

Without telling the employees the purpose of the study, we recently started a survey of those in the older, dark office building. We monitored their sleep, activity, and response to mood through a questionnaire. After a few weeks, they moved to the new building, and we sent out the same questionnaire. We found that in the new building, employees were more active, moved around the office more, and had improved moods. We also found that they slept better at night. The firm was so impressed with the results that they are considering a similar design for their clients.

Today, there is a new trend toward incorporating circadian rhythm science into building design in order to improve the productivity and health of the occupants. Large glass windows are the key to bringing in more daylight, and as the cost of glass goes down and the quality goes up and becomes load bearing and better insulating, workers will be able to enjoy more daylight in indoor spaces.

In many large offices, there is also a recent drive to have open office design with greater ceiling heights to diffuse natural light deep into the workspace. Research labs are experimenting with different office environments to find optimal levels of everything from airflow and temperature to lighting, light composition, and direction of lighting. There will soon be a day when a lighting code for health will be part of a building code.

CHAPTER 7

Syncing Your Exercise to Your Circadian Code

Physical activity is just as important as sleep and good nutrition when it comes to good health. Daily movement improves muscle mass, muscle strength, bone health, motor coordination, metabolism, gut function, heart health, lung capacity, and even enhances how your brain functions. What's more, exercise has a circadian effect, improving sleep and mood. Exercise literally relaxes the brain, reducing depression and anxiety and increasing our ability to experience happiness. Exercise is one of the best medicines. In this chapter, you'll learn how to match your desired physical activity to the right time of day and how to support your activity. Whatever you choose, stick with it: Make physical activity part of your routine so it becomes a habit.

Exercise's mood-boosting effect is critical to keeping you calm and integral to productivity. One of my favorite studies that illustrates this point involves Piet, a milkman from the Netherlands. Piet worked as a milkman all of his life, and he retired around the age of 60. He looked forward to retirement, when he'd be able to sleep in late and stay home to relax. Most of all, he looked forward to spending less time on his bicycle, which he used to deliver milk.

Piet soon developed a new schedule. He would sleep in until 8:00 or 9:00 a.m., and sometimes he would stay in bed until 10:00 or 11:00.

Alone at home, he would stay up late to watch TV, and gradually, his schedule shifted so that he was sleeping as late as 11:00 or noon. He would barely leave the couch, and he snacked on whatever was in the refrigerator. He was also getting frail. After a few months, Piet started feeling low. He started seeing a psychiatrist, but the depression only got worse over time. He finally had to be admitted to a hospital and was scheduled for shock therapy.

A second psychiatrist who worked at the hospital stepped in and read Piet's files. She realized that he had never had any sort of depression during his entire working life, and only small, anecdotal descriptions of sadness from when his sister had passed away or during high school. Initially she thought he might be suffering from postretirement depression, but while Piet was under observation, the psychiatrist noticed changes in his sleeping patterns and sun exposure. She made the connection that in his working life he was waking up early and going outside to distribute milk, exercising the entire time. Yet in retirement, there were whole days when he was staying at home with little exercise or sun exposure.

The psychiatrist changed Piet's sleep schedule and put him in a new room with lots of morning light. She had him meet some of the other people at the hospital, and every day they took walks together in the morning and afternoon. In just a few months, Piet was back to normal. With better sleep, social interaction, and daily outdoor exercise, his depression lifted.

You may be wondering: How could someone like Piet, who was on the verge of getting shock therapy, be brought back to normal by making a few simple adjustments to his daily schedule? And which intervention helped the most? Was it the increased time spent outdoors, the physical activity, or perhaps Piet began eating better, following a schedule, or getting more sleep? We cannot put a finger on any one of these changes being the one thing that improved Piet's condition. We can say that they all played a role in improving his circadian rhythm, and an improved circadian rhythm boosted his health. Let's focus now on the role exercise can play, because for Piet, I believe it was a major factor in his success.

What's Your Minimum Dose of Exercise?

According to the American Heart Association (AHA), anyone healthy enough to exercise needs at least 150 minutes per week of moderate exercise, or 75 minutes per week of vigorous exercise (or a combination of both). That breaks down to moderate exercise for 30 minutes a day, 5 times a week.

Exercise doesn't have to be rigorous or complicated. The AHA believes, as I do, that physical activity can be anything that makes you move your body and burn calories. This includes a vast range of activities, from climbing stairs to playing organized sports.

There are three basic types of physical activity:

- *Aerobic exercise* benefits your heart, is rhythmic in nature, and includes anything that gets your heart rate up that you can sustain for a period of time. *Aerobic* means "with oxygen" and refers to the use of oxygen in the body's metabolic or energy-generating processes; aerobic exercise uses oxygen while engaging large muscle groups.

- *Strength or resistance training* increases muscle mass and overall stamina. This type of exercise consists of short, high-intensity activity and relies on energy sources that are stored in the muscles.

- *Stretching exercises* are best for developing flexibility and proper muscle function (which then supports your strength-training efforts.) Torsten Wiesel, who won a Nobel Prize almost 40 years ago for his work concerning how the brain processes visual information, has remained active and alert into his nineties. Once, while we were hiking in the tropical forest of Costa Rica, I asked Torsten about his secret to a healthy life. He told me that even at 85, he still woke up every morning to do tai chi, as it combined moderate strength training, stretching, and motor coordination. As we get older, we lose motor coordination, and exercise that promotes flexibility and mindfulness helps us stave off that loss.

A Sample MET Table

PHYSICAL ACTIVITY	MET
SEDENTARY LIFESTYLE	<1.5
Sleeping	0.9
Watching television, sitting	1
Writing, desk work, typing	1.5
LIGHT PHYSICAL ACTIVITY	
Walking, 1.7 mph (2.7km/h), level ground, strolling very slow	2.3
Walking, 2.5 mph (4 km/h)	2.9
Gardening, light	2
General house cleaning	2.5
MODERATE INTENSITY ACTIVITIES	3 to 6
Bicycling, stationary, 50 watts, very light effort	3
Walking, briskly, 3.0 mph (4.8 km/h)	3.3
HOME EXERCISE, LIGHT OR MODERATE EFFORT, GENERAL	3.5
Bicycling, <10 mph (16 km/h), leisure, to work or for pleasure	4
Bicycling, stationary, 100 watts, light effort	5.5
Heavy yard work or gardening	4
Dancing (ballet or modern)	4.8
Shoveling snow	6
Mowing the lawn with hand mower	5.5–6.0
VIGOROUS INTENSITY ACTIVITIES	>6
Jogging, general	7
Calisthenics (e.g. push-ups, sit-ups, pull-ups, jumping jacks), heavy, vigorous effort	8
Running/jogging in place	8
Rope jumping	10
Skiing, downhill	6–8
Bicycling, 10–16 mph	6–10
Swimming, crawl, slow	8
Singles tennis	7–12

Relative energy expenditure of various physical activities are described in Metabolic Equivalent of Task (MET). Sitting and doing nothing is often considered as a MET of 1.

The table on page 141 helps you compare different types of exercise over the same period of time, so you can see where you get the most "bang for your buck," metabolically speaking. I find it helpful in deciding which type of exercise to do when I have a limited amount of time. The higher the number on the metabolic equivalent (MET) scale, the more strenuous the activity and the better it is for enhancing your circadian code.

Get Moving!

Irrespective of day or night, whenever you are awake, make sure that you are sitting only when absolutely necessary. Move as much as you can. We expend very little energy while sitting, which has a direct adverse influence on our metabolism, bone strength, and vascular health.[1] When we don't use our muscles, we lose muscle mass and at the same time build up body fat. Being sedentary, even for a few days, can dramatically increase our risk for metabolic diseases, which you'll learn more about in Chapter 10.

The Timeless Benefits of Walking

The simplest and most universal exercise is walking. You can do it anywhere—indoors or outdoors—and you don't need to join a gym. Almost everyone can add more walking to their daily routine. Fitbits and other exercise trackers count steps and suggest that you take a minimum of 10,000 steps (roughly 5 miles) every day to maintain good health and keep weight down. Yet the average health-conscious American who has downloaded a health app walks roughly 4,500 steps each day,[2] while Amish adults in the United States and the Toba hunters of Argentina walk more than 15,000 steps a day.[3,4] What is more astounding is the average user is not increasing their step count, even though they are constantly getting feedback about their daily activity.

While we may not have the luxury of doing the same level of activity as the Toba or Amish populations, we can all make time to exercise and get as close to 10,000 steps as possible.

The Effect of Exercise on Sleep and Circadian Rhythm

Anyone who does a lot of physical activity during the day knows it is relatively easy to fall asleep at night. Even sedentary people who go camping or spend a day in an amusement park report a better night's sleep. We assume that exercise makes us tired. But is there a molecular meaning of being tired? Is there a specific signal from our muscles that tells our brain to fall asleep? Studies show that after exercise, the cells inside our muscles produce several molecules. One of them is interleukin-15 (IL-15), which was already known to increase bone mass. Interestingly, we now know that IL-15 also has some benefits on sleep. In one study, rabbits injected with a small amount of IL-15 were found to have better and deeper sleep.[5]

A second mechanism occurs when muscle cells produce another molecule, irisin. Many obese people have less muscle mass and produce less irisin. Reduced amounts of irisin correlate with obstructive sleep apnea.[6] Exercise for these people can reduce sleep apnea.[7]

While these molecular links indicate the role muscle plays in maintaining good sleep, some new data from mice offers another interesting lead. Mice that lack a circadian clock everywhere in their body and brain have fragmented sleep. But researchers have developed a new genetic method that can turn on specific circadian clocks, like the ones in their muscles. When this happens, these mice sleep like mice that have a clock in their brain.[8] This new finding hints at a completely new mechanism by which muscle clocks regulate the brain and sleep. It implies that nurturing a healthy muscle clock is important for both a healthy body and a healthy mind. Exercise in humans appears to increase the level of an enzyme that is involved in the production of heme—the pigment in our blood that carries oxygen to all tissues.[9] The same pigment is also an important part of the circadian clock, as it tells the clock to turn on and off different genes involved in the metabolism of glucose and fat, as well as the production of hormonelike molecules from the muscle that can go through the bloodstream to affect function of the brain and other organs. This is one of the ways exercise can act on the muscle clock.

I recommend exercise for everyone, and those with sleep disorders might find that it has a strong influence on their circadian code. Even those just starting a new exercise program will see results. They will go to sleep more quickly and wake up less often at night. However, if you have insomnia, see your doctor before you begin a new physical activity program. Insomnia increases your risk of heart disease and stroke, and an exercise program should be done under a doctor's supervision.

The Circadian Component for Maintaining Strength

We discussed how exercise improves sleep and circadian rhythm, but circadian rhythm itself also helps maintain strength so that we are physically fit to exercise. Our physical strength is largely determined by the overall mass and health of our cartilage, bones, and muscles. Each of these key pillars of physical strength has its own circadian clock, which sets a rhythm for repair and rebuilding of these tissues.

Cartilage cells don't have the luxury of multiplying as much as some of the other cells in our body (such as blood cells, liver cells, etc.). But these cells produce the gluelike material that forms the cushion between bones. As we move around, this cushion goes through regular wear and tear. The cartilage cells produce this glue in a daily rhythm, with more produced during the night. When we age or our clock is disrupted, this repair process is diminished,[10] which can lead to osteoarthritis.

With regular wear and tear, bones also go through a daily repair process, which is different from cartilage repair. Our bones are made of minerals, including calcium, that are secreted by cells. Another type of bone cell eats up damaged bone; the circadian clocks in these cells are synchronized so that bone eating and bone making do not occur at the same time of the day. The balance between these two cell types is important. Too much bone eater–cell activity can lead to bone loss, while too much bone making can push against the other bones and create additional damage near the joints. As we age or when our lifestyle is erratic, our circadian clock gets weaker. When this happens, the bone-making cells are not fully activated every day, so they don't produce enough raw materials for making new bone. Similarly, the bone-eating cells are not fully activated, so they don't clear all the damaged bone material completely. This ultimately leads to weaker bones that are prone to fracture. To maintain the healthiest bones, we need to have a strong sleep-wake cycle, eat at the right times, and exercise.

The circadian clock plays a crucial role in both the formation of new muscle fiber and in muscle function. Clock genes directly regulate other genes that are necessary for making new muscle cells or muscle fibers. Clock genes also determine the type of muscles we have. We typically have two types of muscles: Slow-twitch (type I) muscles are rich in mitochondria and help us perform endurance exercise or marathon running; fast-twitch (type II) muscles contain less mitochondria and help us when we are sprinting. Having a better clock appears to increase slow-twitch muscles.[11]

The circadian clock also nurtures our muscles. Depending on whether we just had a meal or have been fasting, the muscle clock activates the function of the metabolic genes involved in absorption and utilization of glucose or fat, which in turn fuels muscle function.[12] The circadian clock instructs other genes to break down damaged muscle proteins and send them to the liver for recycling when we are sleeping. The clock also helps produce new muscle proteins and ensures that the fibers are properly aligned for coherent movement. With all these important roles of the circadian clock in muscle structure and function, it is not surprising that mice that lack a functional clock in their muscles also cannot exercise enough and they get tired too soon.[13]

When to Exercise

Since most of us don't have enough time to exercise, I am often asked if there is an optimal time to exercise in order to reap the most benefits. First, let's talk about duration. If you don't have an uninterrupted 30- to 45-minute block in which to exercise every day, you will experience all the same benefits if you divide your time into two or three segments of 10 to 15 minutes per day. This actually works perfectly for enhancing your circadian code, because exercising during both early morning and late afternoon can boost circadian rhythm. Our ancestors were active throughout their day, but especially in the morning and evening. Many animals in the wild are active at dawn and dusk, so there was a greater need for hunter-gatherer humans to be active around these two parts of the day.

Exercise + TRE = Maximum Fat-Burning Potential

The conventional wisdom is that you should eat before any physical activity. This is not always true. If you have fasted for 10 to 12 hours before your morning walk, run, or bike ride, you will likely tap into your stored body fat for energy during your exercise. If you start your morning activity before breaking your overnight fast, your muscles will spend more energy, using even more fat as the energy source, literally melting away even more body fat. And the more muscle you have, the more calories you'll burn throughout the day and the leaner and healthier you'll be. While engaging in strenuous strength exercise or physically demanding sports such as rowing, soccer, or basketball in the morning with an empty stomach may not be ideal for peak performance, a morning walk, moderate running, or road biking are tolerable physical activities before breakfast.

Exercise in the Morning

Early morning is a great time to get outside and start moving with an aerobic activity. A brisk walk, or any outdoor activity in the presence of bright daylight, is an excellent way to synchronize the brain clock, making it a way to beat any form of jet lag or help you recover from sleep deprivation. It is also an important mechanism for maintaining and enhancing brain function. First, it will improve your mood for the rest of the day. Also, exercise stimulates new brain cell production[14] and your ability to make new neuronal connections for deeper learning and more memory. We also know that exercise helps repair damaged brain cells by improving the neurons' ability to repair their own DNA.[15] This damage repair extends to the plaques in the brain that are found in those who have Alzheimer's.[16]

It doesn't matter whether you wait until sunrise to start your morning walk, run, swim, or cycle. You can start anywhere from 30 minutes to 2 hours before or after sunrise. The outdoor light during this time can be as much as 800 to 1,000 lux, which is an ideal amount of comfortable daylight. This bright light will activate the blue light sensors in your eyes, and as you exercise, the sensors will fire up your brain. If you exercise in a gym in the morning, don't choose the darkest corner of the room. Instead, find a spot that is next to a large glass window or under bright light.

As long as you are dressed properly for the weather, you can take a morning walk throughout most of the year, unless there is a weather advisory. In fact, exercising in cold air imparts some additional health benefits. Cold air activates brown fat or converts white fat to beige fat.[17] Brown fat is rich in mitochondria, the energy currency of any cell. More mitochondria mean that fat cells have more capacity to burn off. Additionally, body fat is burned to warm up the body during cold-air workouts. As a result, you can simply burn some fat by being exposed to cold temperature.[18]

Early Morning Outdoor Exercise Benefits

Outdoor physical activity early in the morning is ideal for many reasons.

- You get some daylight exposure to sync up your brain clock.

- Exposure to daylight increases alertness and reduces depression.

- On colder days, you activate brown fat and increase your fat-burning potential.

- You naturally raise your cortisol to a healthy level in the morning, which will lower inflammation.

Exercise in Late Afternoon

Another great time for physical activity is at dusk or in the late afternoon,[19] starting from 3:00 p.m. to dinnertime. This is when muscle tone begins to rise, so it's the best time for strength training, including weight lifting, or vigorous exercise like intense indoor cycling. High-intensity athletes and those trying to optimize their physical fitness will find that exercising before dinner followed by a protein-rich meal will help them repair muscle, build muscle mass, and promote recovery.

The circadian component to this aspect of peak performance may come from various internal clocks. The muscles absorb and use nutrients in the late afternoon, when repair occurs. The brain's function involving motor coordination is typically high during the day, which further aids sports performance. Blood flow and blood pressure are also high in the afternoon, which might improve better oxygenation of muscles.

There is also a circadian rhythm for exercise performance. Athletic performance, even among competitive athletes, can vary by as much as 25 percent within a day.[20] If you are aiming to get maximum benefit from exercise with minimum injury, afternoon is the best time to exercise. There are numerous studies showing motor coordination and strength peaks around late afternoon. This is further supported by observation from analysis of 25 seasons of Monday Night Football games, dating from 1970 to 1994.[21] When a West Coast team traveled to the East Coast to play Monday Night Football within 48 hours after flying, the West Coast team had a significantly higher chance of beating the East Coast team, even though the East Coast team had the home-field advantage. That is because the East Coast team started the game at 9:00 p.m., at the tail end of their peak exercise performance window. However, the West Coast team was still following their old time zone's circadian clock and actually playing at their peak performance time of 6:00 p.m.

For average mortals (the majority of us), late-afternoon or evening exercise has two practical benefits. Exercise is known to reduce appetite,[22] so afternoon exercise not only helps burn some calories, it can also help reduce hunger at dinnertime, so you may eat less. Exercise also helps our muscles take up more glucose in a mechanism that does not depend on insulin.[23] As insulin production and release gradually decline through the evening, insulin alone may not be sufficient to prevent our blood glucose levels from shooting up beyond the healthy range. As little as 15 minutes of evening exercise will boost our muscles' ability to absorb some blood glucose and keep it in healthy range.

Some people worry that if they push their intense exercise to the outer limits of this window they will not have enough time between dinner and sleep. Let's say you're working a traditional 9:00 a.m. to 5:00 p.m. job and you exercise after work and then you have dinner. Now you're pushing dinner out until 7:30, 8:00 p.m. That's okay because exercise absolves some sins: The positive benefits of exercise outweigh a lost hour or two of TRE. If you are going to do endurance exercise, where you are trying to push your limit and go that extra mile, then you've got to remember that you're only going to do a 10-hour TRE, and that's ok.

Better than Nothing: Exercise after Dinner

If you cannot exercise in the morning or afternoon, evening exercise is better than nothing, and it has its own set of specific benefits that affect your circadian code for metabolism and maintaining blood sugar levels. Physical activity increases demand for glucose, and muscles can soak up a good amount of blood glucose, thereby reducing the blood glucose spike after an evening meal so that you will be in a normal physiological range. After dinner, mild physical activity, like an evening walk or doing chores in the house, also helps digestion by moving the food down the digestive tract and reducing the chance of acid reflux or heartburn. Since the insulin release and subsequent action on blood glucose regulation declines in the evening,[24,25] for those at risk of type 2 diabetes, any physical activity in the evening is like taking a diabetes pill to reduce blood sugar.

We don't know exactly if exercise after dinner affects your sleep, but we do know that any physical activity promotes better sleep. And we do know that bright light exposure at night can delay your sleep time. If you have to get your exercise time in after dinner, it's better to do it away from bright lights.

Yet not all exercise at night is a good idea. It's best to do your extreme activity or high-intensity exercise before dinner. Late-night exercise in a gym or on a treadmill can increase cortisol to morning levels and delay the nightly rise of melatonin. Intense exercise also raises body temperature and heart rate. All of these factors interfere with your ability to go to sleep. You may be resetting your clock by sending a signal that it's earlier in the day. What's more, if you do very intense exercise at night, the brain thinks it is dusk, when we are typically more active, so it delays melatonin production. This may be a reason why some (not all) people who exercise late at night also go to bed after midnight. If late night is the only time you can exercise, taking a shower before bedtime can help your body cool down, which will help you get to sleep.

When Should Night-Shift Workers Exercise?

Night-shift work often involves physical activity, so many night workers may not need extra activity. But the nature of night-shift work in many professions has changed and become more sedentary, which can make workers sleepy and trigger a reliance on caffeine to keep them awake, which in turn interferes with their attempted sleep once they get home.

Although there is not too much scientific data to support whether exercise can be a timing cue to reset our circadian clock to a new time zone, we do know that exercise can reset the circadian clocks throughout our body and brain. As nighttime exercise can increase alertness and suppress sleep, it can be used to the advantage of night-shift workers. In fact, Cory Mapstone, a veteran police sergeant with the San Diego Police Department, has figured out how to align his circadian code with his shiftwork. On quiet nights during his shift, he drives to a community park and takes a few minutes to get in some high-intensity exercise that increases cortisol production—a few push-ups, jumping jacks, lunges, etc. Cory reports that this trick has helped him avoid the trap of coffee and energy drinks, ensuring that he gets good sleep once his shift is over.

Timing Meals Improves Exercise Performance

Just like exercise improves sleep and circadian rhythm, having good sleep and circadian rhythm also has rewards on exercise performance. It is well known that getting a good night's sleep is a necessity for optimum athletic performance.[26] But what about diet and timing of meals?

High-intensity athletes are known to eat a lot of protein to build up their muscles. Unless you are going to compete in the Olympics, stick with the balanced diet described in Chapter 5. Everyone should focus more on when they eat, instead of what they eat. Our research found three substantial benefits related to diet and exercise when we fed mice for only 8 to 10 hours compared to letting them decide when to eat. The first improvement was in muscle mass. We assumed that fasting for 14 to 16 hours would break down muscles and we would see a reduction in muscle mass. Actually, we found the exact opposite. When mice ate for only 12 hours, we never saw a reduction in muscle mass. In fact, only fat mass was reduced. If the mice ate a healthy diet within 8 to 10 hours, they gradually increased their muscle mass, and after 36 weeks, they had 10 to 15 percent more muscle mass than mice that ate whenever they wanted.[27]

We also know that many genes that are involved in repairing muscle and growing muscle are circadian, with peak production during the day. These genes are directly under the instruction of both a circadian clock and the feeding-fasting cycle. In our lab, we found that muscle repair and rejuvenation genes in the mice got a double boost from having a healthy circadian clock and also having a clear feeding-fasting cycle. This might explain why they gained more muscle mass.

We have not yet directly tested this on athletes. But there is some anecdotal evidence that shows it may be true. Several personal trainers are now adopting an 8-hour eating window along with exercise as a bodybuilding recipe. Hugh Jackman's famous Wolverine diet is actually an 8-hour TRE interval. A systematic study of resistance-trained athletes on an 8-hour TRE also showed some benefits.[28] Remember, these were resistance-trained athletes with excellent physique and body composition to begin with. They were already paying extreme attention to every ounce of fat and muscle mass they have. So, researchers were not expecting too much additional benefit from adopting a 10-week TRE. These athletes did not see a decline in muscle mass, but surprisingly, their fat mass significantly reduced, and many markers of good health also improved. This has led us to believe that for the entire range of people, from nonathletes to extreme athletes, TRE offers health benefits.

The second improvement we saw in the mice who were subjected to TRE was an increase in endurance exercise capacity. Running a marathon is a lot of stress on the body and mind. Enduring that pain, and in fact enjoying it, is a marker of resilience. When we start such long physical activity, our body initially taps into the readily available sugar as an energy source, and when glucose or glycogen is depleted, we "hit the wall." Our brain and body get exhausted of energy and we cannot run any longer. Endurance training helps muscle do two hugely beneficial metabolic adaptations: Muscle learns to absorb more glucose from blood when food is available so that there is a larger store of glucose and glycogen to use during endurance training; it also learns to adapt to an alternate energy source when all stored glycogen is used up. When this happens, muscle switches to using stored fat as its energy source. The fat is converted to ketone bodies, and this simple carbon source is used as fuel for the extra miles.

Combining time-restricted eating with endurance sports gives our body double benefits. TRE boosts the signals for muscle repair and regeneration to help sustain or build muscle mass, and increased physical activity helps soak up more glucose from the bloodstream into muscle, so the extra glucose is diverted from the liver, where it was to be stored as fat (which could lead to fatty liver disease).

The third improvement in the mice links motor coordination and

Rhonda's Exercise Performance Increased with TRE

Rhonda Patrick has a podcast called *FoundMyFitness*, and I was lucky enough to be asked to appear on her show. Rhonda is very careful about dieting and exercise. When she started doing a 12-hour TRE, she felt great, and she told me that she was more alert and that her subjective sense of health increased. When she tried a 10-hour TRE, she realized that her endurance went up. She felt less tired after miles of running or biking. However, when she went back to a 12-hour TRE, that benefit went away.

The sweet spot for creating an endurance boost can be different for each of us. We also don't know if the type of diet one eats can impact this sweet spot. In our TRE experiments with mice, we found a modest increase in ketone bodies in mice that eat for only 8 or 9 hours and fast for 15 or 16 hours every day. Ketone bodies are known to be produced after several hours of fasting. An increase in ketone bodies is linked to an increase in endurance.[29] Rhonda perhaps experienced a modest rise in ketones when she did the 10-hour TRE but not when she was on the 12-hour TRE. A diet that is rich in fat or ketone bodies can naturally boost ketone production, while a diet rich in carbohydrates may not. Therefore, people may watch their diet and how many hours of TRE they are doing to find their own sweet spot of improved endurance.

TRE. We found that the mice following a restricted eating pattern had increased motor coordination. In my lab, we put mice in a rotating drum, where they have to balance themselves. We found that if they eat for 8 to 10 hours, they can stay on the drum 20 percent longer. Motor coordination is important throughout our life span, but it's especially essential as we age.

Is 8 hours the magic number in which to squeeze all your calorie intake? We don't know this for sure, yet hundreds of athletes and health enthusiasts who use our myCircadianClock app or self-monitor their eating pattern and try to determine how long to bike or run on the treadmill or trail tell us somewhere between 8 and 10 hours is the sweet spot to improve endurance. When they eat for more than 10 hours, most of them lose the extra advantage of time-restricted eating on endurance without losing other benefits of better sleep or reduced fat.

People who exercise on a regular basis report they feel less hungry the rest of the day,[30] making an 8-hour TRE more manageable. The reason is that exercise reduces your hunger hormone ghrelin and increases satiety hormones, which are also under circadian control. Intense exercise has a stronger effect on hunger compared to moderate exercise. However, you have to keep up the exercise habit, because this benefit wears off within a few days.

CHAPTER 8

Adapting to the Ultimate Disrupters: Lights and Screens

Modern life has meant less access to natural light during the day and more artificial light at night for at least 100 years, yet industrialization and electricity is not what has finally pushed us toward an almost circadian collapse. Instead, it was the sudden and universal ubiquity of digital screens. While shift work was the primary disrupter of circadian rhythm only a few years ago, today, connectivity is the culprit.

We live in a time-displaced world, controlled by a 24-7 news and entertainment cycle. The virtual world has no day or night: We can always find someone to chat with, or something to entertain us or fill the gaps or sleeplessness or boredom. And when we aren't glued to the latest cat video, celebrity meme, or natural/political disaster, we are trying to stay in touch via our social networks with friends, family, or co-workers who often live in different time zones. This lifestyle has created an entirely new type of circadian disruption—a digital jet lag—where our body is in one location, but our mind is operating in another.

Yet we know that the physical body is not designed to be in a constant state of wakefulness. When cancer experts say that shift work is a known carcinogen, they are referring to bright light exposure that enables shift workers to stay up at night. A recent report from the National Toxicology Program evaluated the non–cancer related health problems related to light. They found that light exposure at night may be linked to heart disease, metabolic disease, reproductive issues, gastrointestinal disease, immunological disease, and a number of psychiatric diseases.[1] Interestingly, these are the same chronic illnesses that many Americans face, and they are known to have a circadian component. In Part III, we'll discuss each of these problems individually.

In terms of connecting the effect of bright light in the middle of the night to the circadian code, we know that it can cause the whole circadian rhythm to collapse. Charles Czeisler of Harvard University did a simple experiment in the 1980s. He took healthy volunteers and recorded their body temperatures. Then he exposed them to a bright light at different times of the night. The next day, he recorded their body temperatures and found that among the volunteers who were exposed to bright light between midnight and 2:00 a.m., the circadian rhythm for core body temperature completely collapsed the next day, as if their bodies instantly lost track of time.[2] Restoring their normal light-dark cycle was all that was needed to regulate their body temperatures the third day.

Some experiments with mice indicate the effect of light might extend beyond simple alertness, sleep, depression, and migraine to more serious cases of seizure and epilepsy. A certain type of epilepsy, called *nocturnal frontal lobe epilepsy,* usually occurs at night, although some forms of this disease are triggered by bright strobe lights at any time of day. In humans, the disease is caused by a mutation in a gene called *cholinergic receptor nicotinic beta 2* (CHRNB2). One of my esteemed colleagues at the Salk Institute, Steve Heinemann—famous for discovering several molecules of the nervous system—studied mice that had the same human mutation that causes nighttime epilepsy. However, the mice never showed any sign of epilepsy, and Steve lost interest. It does happen sometimes—a disease in humans that cannot be exactly replicated in mice or vice versa. I had some interest in the gene, as it showed circadian rhythm in the brain, so I thought it might be involved in arousal and sleep regulation. When we monitored the circadian activity pattern of these mice, we realized that they actually had a sleep problem. While normal mice wake up in the evening and stay active through the night before falling asleep in the morning, the beta 2 mutant mice woke up in the middle of the night and were active well past morning,[3] as if their normal response to light was altered. Interestingly, patients with nocturnal frontal lobe epilepsy also stay awake late into the night and are very sleepy throughout the day. Although the mice could not replicate the seizure phenotype of humans, we were satisfied that the sleep-wake pattern of the mutant mice was a mirror image of that in human patients. These experiments gave us the initial clue that the gene may function by cranking up or down the light signal that goes from the eye to the brain, to help the brain decide whether to stay awake or to sleep.

A few years later, Marla Feller of the University of California, Berkeley, found another surprising result. She noted that mice that lacked this gene were supersensitive to light in the blue light spectrum. Even under dim light, the nerve cells in their eyes would fire up as if their eyes were exposed to very bright light.[4] The defect was traced back to excessive sensitivity of the melanopsin cells to light. Early in life, our eyes are not completely wired to our brain. The ganglion cells from the eyes that transmit all light information from the eye to the brain actually branch out or are dedicated to connecting to numerous brain regions to regulate how light affects vision, behavior, sleep, alertness, depression, seizure, migraine, etc. This patterning of the ganglion cells is heavily studied. As you can imagine, a misconnection between the eye and the brain can have lifelong consequences. What is surprising is that although melanopsin is present in only a small subset of 2 to 4 percent of all ganglion cells, when these melanopsin cells are less active or more active, they also affect how the rest of the 96 to 98 percent of cells wire to their respective brain targets. The mice lacking the beta 2 gene had more sensitive melanopsin ganglion cells and their overall connection of ganglion cells to the brain was also defective.

Conversely, David Berson of Brown University showed that mice lacking the melanopsin gene also had a defective connection to the brain.[5] These experiments in mice predicted that several neurological

Different Types of Light That Are Rich or Poor Sources of Blue Light

Light Color Composition

Light Source	Color Temperature	Violet	Indigo	Blue	Green	Yellow	Orange	Red	
Daylight	5500–7500 K	+	+	+	+	+	+	+	• Promotes alertness • Reduces sleep • Best for daytime • Disturbs circadian rhythm at night
Cool white LED	6000 K	-	+	+	+	+	+	+	
Computer/phone screens	6500–7500 K	-	+	+	+	+	+	+	
Natural white LED	3000–4000 K	-	+	+	+	+	+	+	
Warm white LED	4000–5000 K	-	+	+	+	+	+	+	
Compact fluorescent	6000 K	-	+	+	+	-	+	-	
Incandescent bulb	2700 K	-	-	-	+	+	+	+	• Insufficient for daytime activities • Best for evening • Task lighting • Less harmful for maintaining circadian rhythm at night
Halogen bulb	3000 K	-	-	-	+	+	+	+	
Outdoor/high pressure sodium	2200 K	-	-	-	-	+	+	+	
Candlelight style OLED	2000 K	-	-	-	+	+	+	+	
Candle	1800 K	-	-	-	-	-	+	+	

diseases in humans, including migraine, epilepsy, seizure, and even excessive light sensitivity, can have an underlying problem with how our eye is connected to the brain. While these are very debilitating diseases with underlying disease-causing mutations, less-severe forms of these mutations may not cause disease but can have very subtle effects on our lifelong sensitivity to light. Some people may be less sensitive and have no trouble falling asleep under usual light in a living room, whereas others may find that the same level of light keeps them awake late into the night and that they can sleep only in a darkened bedroom.

Even dim light can interfere with your circadian rhythm. A mere 8 lux—a level of brightness exceeded by most table lamps and about twice that of a night-light—has an effect, notes Steven Lockley, a sleep researcher at Harvard. Staring at most screens at mid- to high brightness introduces more blue light to our retina and brain. The blue wavelengths—which are beneficial during daylight hours because they boost attention, reaction times, and mood—seem to be the most disruptive at night. Exposure to them reduces melatonin production and suppresses sleep. For children and teens, screens filled with blue light pose a particular problem. A 2016 study on 600 children showed that children who have increased screen time are more likely to have poor sleep quality and problem behaviors.[6]

Reducing Blue Lights on Our Screens

Just like controlled use of fire revolutionized human life, judicious access to the digital world holds the key to regaining our health. As we spend more than 8 hours a day looking at digital screens, screen brightness and color is a significant source of light exposure.[7] Reducing blue light from screens is a smart approach to reducing evening exposure to blue light.

It is truly gratifying to see how the basic discovery of melanopsin in frog skin in 1998[8] has transformed into a blue light revolution. For instance, Michael Herf—inventor of the famous photo-editing software Picasa, which eventually merged with Google Photos—took a special interest in our research. He recognized that a simple app that changes the brightness and color of a traditional blue light screen to a slightly orange-looking screen that is low in blue light may help some people. He devised the f.lux app, which can be downloaded to any PC or Android phone. It can be programmed to automatically change the color and brightness of the screen to more soothing and less disruptive shades of orange or red that match with the user's preferred sleep time. Thousands of people around the world have already downloaded the app, and clinical studies have shown that reducing blue light exposure through f.lux improves sleep and reduces eye fatigue.

Seeing the wild success of such a simple app, Apple, Samsung, and other phone manufacturers have made it a standard feature of many smartphones. Apple calls this their Nightshift feature: All you need to do is provide the time for 2 hours before your preferred sleep and wake times, and the app takes care of the rest, reducing the blue light of the screen to transform it from a bright white to a beige glow. Almost every new laptop and tablet that's coming to market now has a built-in function to set the time at which the brightness or screen color will change.

It's very gratifying to see how our discovery in mice went from simple observation to adoption on more than a billion devices within 15 years.

Many new televisions also include this technology. Features such as Samsung's Eye Saver Mode gradually change color and reduce blue light on TV screens. Your eyes will adapt slowly so that you won't even notice the change in color as it's happening during your favorite show. That way you can enjoy your TV without having your sleep affected by blue light.

If you don't want to run out and buy a new television, add-on products can transform your current one. For example, Drift TV is a small box that connects to your television through an HDMI input and removes a percentage of blue light from the screen. You can set how much blue you want to take out: For example, you can set your Drift TV to remove 50 percent (or any percentage in increments of 10) of all blue light over a period of 1 hour. That way, the transition is seamless and virtually unnoticeable.

Home Lighting Is an Easy Fix

Our discovery in lighting and blue light has inspired lighting manufacturers, architects, lighting engineers, and interior designers to rethink indoor lighting. These professionals are pinpointing circadian lighting as the next big financial opportunity, so this area is ripe for further innovation and competition to bring circadian lighting into our homes.

The continuing evolution of the light bulb creates new challenges and opportunities for circadian-rhythm restoration. For instance, LED (light-emitting diode) bulbs were first produced in the red and green spectrums, as blue-spectrum lights were difficult to produce for a long time. Recently, the 2014 Nobel Prize–winning work of Isamu Akasaki, Hiroshi Amano, and Shuji Nakamura made blue-spectrum LED bulbs more affordable. The amount of light produced from these blue LED bulbs increased severalfold, such that a 12-watt LED bulb is now as bright as a 60-watt light source of 10 years ago. This seminal invention in lighting reduced power consumption and led to the industrial-scale production of LED lights. While these LED lights are much more energy-efficient than the old-fashioned incandescent light bulbs, they also produce more blue light, which disrupts your ability to fall asleep at night. As more people make the switch from incandescent bulbs to these cheaper LEDs, society's circadian problem is only getting worse.

Some of our greatest home-use products and scientific advances have been popularized by NASA: Velcro and Tang, for example. We've learned from the International Space Station that astronauts experience severe circadian disruption. They lose the sense of day and night due to the constant lighting and lack of connection with true sunrise and sunset. To improve their sleep and circadian rhythms, NASA is changing the light bulbs in the space station to new LED bulbs that change color and dim.

These tunable LED lights are available for home use as well, and the brightness and color of the light can even be adjusted from a smartphone or a remote control. They can also be programmed to change color and brightness at different times of the day. In other words, we can re-create seminatural lighting by increasing blue light during the day and amber-colored light at night, simulating a natural day-night cycle. The lights brighten in the morning from complete darkness to bright blue; toward the end of the day, they'll slowly dim down to an orange glow and then to complete darkness. The cost of these tunable light bulbs is high right now, but as lighting trends from the past 100 years have shown, the cost may soon come down. The bulbs are available online as well as at many hardware stores.

For now, home owners can install dimmer switches on their current LED lights. In the daytime, the lights can be set at full blast, and at night they can be dimmed to a level that is just bright enough to allow for safe movement around the home. Another easy fix is to put different light bulbs in different rooms. For example, if you have two bathrooms, install dim lighting in the bathroom you typically use in the evening and bright blue-enriched LED lighting in the bathroom that you use in the morning. As you wake up and walk into the bathroom with bright blue light, the exposure to the light will start to reduce your melatonin and make you feel more alert.

If you wake up frequently at night to use the bathroom, you can install motion-activated path lighting that shines directly on the floor. This type of lighting is the least disruptive and will not activate the blue light sensors in your eyes. I've found that it is now a standard feature in many hotels, and I can imagine how helpful it might be in hospitals and nursing homes as well as on the consumer market.

Or you can change your light bulbs to amber-colored lights that throw off a bit of orange color. These bulbs have less blue light and won't disturb your circadian clock as much. They also support the rise of melatonin in the evening, so that everyone in the home will feel sleepy around 10:00 or 11:00 p.m. Lighting departments at many major home retailers will have samples so that you can clearly see the difference.

You can also adjust the type of lighting you use in the evening. For reading, or for doing homework, you may need a little bit more light than what is provided by your dimmed overhead lighting. Instead of flooding the room with light, focus on task lighting from table lamps. This type of light actually falls on the work surface and not on your eyes, so you can still work under relatively bright light but with less total exposure.

Red lights have the least amount of blue and work well for night-lights. For instance, there is a television show in the United Kingdom

Teenagers, Light, and Computers

We have done some preliminary research that shows that adolescent boys like living in the dark. This is clearly disruptive to their circadian rhythms: During the day when they are supposed to be exposed to bright light they avoid it, and then they spend the nights looking at screens in a dark room. So, if you are a parent of a boy and see this behavior, encourage him to open the shades, and to program his computer and phone to emit less blue light at night 2 hours before bedtime.

called *Doctor in the House*. The host, Dr. Rangan Chatterjee, moves in with families temporarily to tackle their health problems. He analyzes their lifestyle to see what simple changes they can make to improve their health. He follows my work very closely, and one of the recommendations he makes is to change the night-lights in children's bedrooms to red lights. When he does this, he has found that he can extend the child's sleep time by a full hour.

Try Blue-Light-Filtering Eyeglasses

We have known for more than 30 years that blue-light-filtering glasses relieve chronic migraine pain. In a study done in the late 1980s—well before the blue light effect was understood at molecular and neural levels, a physician had a suspicion about the color of light affecting migraine pain. He did a simple experiment with children who were missing school days every month due to migraines. The kids were divided into two groups. One group got to wear blue-light-filtering pink glasses and the other got orange-light-filtering blue glasses. The kids with blue-light-filtering glasses had fewer incidences of migraine and the attacks were also of shorter duration, and they missed fewer school days.[9]

In 2010, Kazuo Tsubota, professor of ophthalmology at Keio University Graduate School of Medicine in Tokyo, heard about our work on blue-light-sensing melanopsin cells. He had been witnessing a disturbing circadian disruption trend in Japan. Young kids were spending too much time in front of computer screens or playing video games, getting very little sleep at night, and feeling tired all day. Older people were also spending too much time staying awake watching late-night TV. Japan, being the leader in lighting technology, was also rapidly adapting LED lights. Dr. Tsubota determined that it would be a losing proposition to convince people to dim their lights. Instead, he came up with a simple idea: Blue-light-filtering eyeglasses might be extremely helpful for reducing eye stress and improving sleep. In the same way that you wear sunglasses during the daytime to protect your eyes from direct sunlight,

The Trickle-Down Effect of Scientific Discovery

In 2013, Dr. Tsubota convened a blue light conference in Tokyo. It was the first time that lighting engineers, ophthalmologists, psychiatrists, and scientists like me got together and discussed how to manage the new wave of LED lighting. What Dr. Tsubota triggered in Japan just a few years ago is now reverberating throughout the rest of the world. In March 2017, while attending the Near Future conference, which brings together thought leaders from various fields, someone approached me to sell me blue-light-filtering "melanopsin glasses" for $99. A month later when I went to my optometrist for a new pair of glasses, she asked me if I needed a blue-light-filtering coating on my new prescription glasses.

his blue-light-filtering glasses are meant to be worn in the evening to reduce the amount of blue light hitting your eyes when you are at home watching television or out in a supermarket, drugstore, or gym.

By the time I met Dr. Tsubota in 2012, he had already custom-designed a pair of glasses with what looked like pink shades. Every evening around 7:00 p.m., he would remove his normal day glasses and put on the pink ones. He personally experienced a better night's sleep. He then convinced an eyeglass manufacturing company—JINS—to come up with a consumer product that could be priced under $25. JINS' blue-light-filtering glasses sold like hotcakes in Japan, and now many additional eyeglass manufacturers sell them through optometry stores in the United States or on the Internet. Now, even "transition" lenses that change from clear to dark to protect wearers' eyes from the sun are being marketed as "blue-light-filtering" glasses.

You can start wearing blue-light-filtering glasses right after dinner, and within 10 to 15 minutes, your eyes will relax, you'll experience less eye strain, and your brain will adjust to the color. People may think you are a huge Bono fan, but that's okay; at least you are in control of the light that goes into your retina.

If you wear blue-light-filtering glasses, then you don't have to change the light bulbs in your home or find apps for your laptop or television. However, if you wear prescription eyeglasses, do not coat the glasses you wear during the day with a blue filter, because you still need blue light during the daytime. (If you're traveling, it will actually make your jet lag worse.) If you want to adopt blue-light-filtering glasses, make sure that you have a separate pair of glasses only for the evening. Put them on only for 3 to 4 hours before going to bed.

Lastly, pay attention to the color of the lenses. The orange/pink–hue filters out the most blue light; other colors filter out only 5 to 15 percent of blue light, too small to make a real difference.

Robert Wore His Blue-Light-Filtering Glasses During the Day

While it is gratifying to see how basic scientific discoveries from my lab are transforming lives, I am also getting a little worried. In April 2017, I got a phone call from my friend Julie Wei-Shatzel, a primary care physician in Folsom, California. Dr. Wei-Shatzel told me about one of her patients who was experiencing severe jet lag and depression-like symptoms after a recent trip from the East Coast. She figured out that her patient, Robert, had just bought a new pair of prescription glasses with a blue-light-filtering coating; they were sold to him as a product to wear while he worked at his computer. The glasses were meant to reduce eye strain. While they were effective, Dr. Wei-Shatzel found that continuous use of blue-light-filtering glasses during the day essentially filters out most of the blue light we badly need to maintain mood and entrain the circadian clock to the local time. With less blue light coming through his glasses, Robert's clock was stuck. The lack of bright light mimicked a northern Canada winter, and his brain was slowly drifting toward depression.

Dr. Wei-Shatzel was aware of my team's work on blue light, so she asked Robert to switch back to his older glasses and see if it improved his mood. In a couple of weeks, he was back to normal, with a better mood and no jet lag.

Another Form of Jet Lag: Lighting in Hospitals

Managing light is becoming more important in hospital settings, where maintaining a robust circadian rhythm can substantially improve recovery and healing. Most hospital rooms are lit as if the patient is living under continuous twilight. This is even more serious in the neonatal intensive care unit (NICU), where premature babies without a well-developed circadian clock spend weeks under almost constant light. In an interesting study, yet to be replicated in additional hospitals, just covering the cribs in the NICU with a blanket for a few hours to create a sense of night substantially improved the health of these very fragile babies, who were then released from the NICU to normal care much faster than those who stayed under the standard care in constant light.[10]

Measuring Light on Your Own

Light is an interesting environmental factor that plays weird tricks with our brain. When you step from indoors to outside on a bright sunny day, you are initially blinded by the glare and brightness. But within a few minutes, you can completely adjust to the bright daylight and be functional. Conversely, when you step into a dark movie theater, you have difficulty finding your way, but within a few minutes your brain gets used to the darkness and you can see things that you did not see earlier. Hence, it is tricky to rely on your eyes and brain to assess the brightness of a room so that you can figure out how much light you need or need to avoid.

In circadian rhythm research, we routinely use a watchlike device that measures our movements and figures out step counts and total minutes of sleep. Many of these devices also sense light every 30 seconds over the course of several days. I have been wearing one of these watches for a few years. When I looked at my light-exposure pattern while camping in the Maasai Mara National Reserve in Kenya, my watch told me I spent more than 8 hours a day getting 2,000 lux or more of light, even though I spent most of my time inside a pickup truck, under a tree, or in a tent. A few days later, when I was working in a lab in Nairobi with good-size windows that brought in plenty of light, I was still getting 2 to 3 hours of bright light above 2,000 lux and many hours of diffuse daylight at 300 to 500 lux. A few days later, back in San Diego in my home and office, I was surprised that my light readings during the weekdays were miserable. I was hardly getting an hour of bright light, and most of it was when I was driving between home and work.

In my lab, we have since looked at light exposure from hundreds of people living in "sunny San Diego" using the same wrist-worn devices. Most of them get their daylight when they drive, or when they sit outside for coffee or food or take a walk. These wrist readings were also misleading, because even though the light was hitting their wrist, many people were wearing sunglasses, which can reduce light reaching the eyes by seven- to even fifteenfold.

Not everyone has access to a fancy light meter in their watch (at least not yet, but we hope that some activity trackers or smartwatches will add a light sensor soon). Yet having one would be beneficial. For example, some people tend to feel tired and sleepy in the evening, but when they go out to run some errands—getting groceries, milk, or beer from a corner store or going to the pharmacy or simply walking in a mall—after a few minutes they are alert all over again. This may be connected to the indoor lighting. The average grocery store, pharmacy, gas station food mart, or store at a mall has at least 500 lux of bright light. Some stores even light up their shelves, and that light hits the eye horizontally. This amount of light is several hundred times brighter than what our brain is designed to experience in the evening, when it is trying to calm down and go to sleep. So, it is no wonder that we feel fired up after a trip to the grocery store in the evening.

A few years ago, Ben Lawson, a high school student working in my lab, figured out how to use the camera in his smartphone to measure light. His app, myLuxRecorder—now available for free—can be used on iPhones to get a good light reading anywhere. This has helped me figure out how bright certain stores can be. You can do this same experiment and survey your nighttime exposure and limit it where you can.

What About Sunglasses?

Sunglasses can reduce bright light reaching the eye by seven- to fifteenfold. That means if daylight inside a car is around 5,000 lux, sunglasses cut the exposure down to between 330 and 700 lux. Thinking about this math and knowing that my major source of daylight is when I drive to and from work immediately made me quit wearing sunglasses during my regular day's activities.

You may think that the UV rays from the sun can damage your retina. But in reality, most people who work inside an office—like me—are barely exposed to direct sunlight for more than a few minutes every day. Our car windows and windshield, as well as the cornea and the lenses of the eye, actually filter out a lot of UV light before it can cause damage.

Even in California, I only wear sunglasses when we take a road trip or spend a few hours on the beach. On regular days, when I spend less than an hour driving, I don't wear them and choose instead to get full daylight exposure to help set my circadian code. And of course, I never look directly at the sun.

Technology to Keep Us on Track

Tracking our daily rhythms can help us more clearly evaluate how our eating, sleeping, and activity patterns may be helping or hindering our natural circadian code. But what to monitor? There are many parameters that can be monitored by consumer-grade technology, while some medical-grade devices might be useful as well. For instance, our heart rate, blood pressure, and body temperature have daily rhythms that are supposed to dip in the evening and begin to rise in preparation for waking up. If you had access to these patterns on a regular basis, you could tell how closely aligned you were to an ideal circadian rhythm and then make any necessary adjustments in real time. A nighttime dip in blood pressure is a good measurement of a healthy heart. Similarly, a daily rhythm in core body temperature is an indicator of a strong circadian rhythm.

A body-surface-temperature reader, which will be found on upcoming versions of wearable technology, will mirror core-body-temperature rhythms and should show an increase in surface temperature at night and a small dip during the day.

Body Temperature and Ovulation

Our body temperature has a predictable 24-hour rhythm. Women of reproductive age also have a superimposed temperature rhythm that coincides with their menstrual cycle. Consumer-grade vaginal temperature biosensors can continuously measure temperature every 5 minutes for several days and can predict the exact time when women are fertile and infertile.[11] Such fertility awareness can help plan pregnancy with higher accuracy.

Another marker of health is bloodwork that measures peripheral capillary oxygen saturation (SpO2). When we sleep, our SpO2 level should be maintained at more than 95 percent, but some people with severe sleep apnea may see numbers that drop below 95 percent. Therefore, monitoring these rhythms with a home dissolved-oxygen monitor can offer a glance into the body's oxygen rhythm.

There is growing interest in using the data from wearable sensors to monitor our circadian rhythm, and there are a number of scientific papers that have assessed the utility of consumer-grade wearable sensors.[12,13] We hope the technology will soon be available to measure our own internal rhythms and follow how changing sleep time, exercise, or eating habits bolsters or dampens our rhythms.

While the above rhythms can be measured from the skin without poking a needle, a continuous glucose monitoring system (CGMS) inserts a hair-thin sensor into the skin and can measure interstitial blood sugar continuously in 1- or 5-minute intervals for 7 to 14 days. This is an exciting technology currently prescribed for people with diabetes. Eran Elinav, a professor at the Weizmann Institute of Science in Israel, has used these glucose monitors on dozens of healthy individuals, who were then instructed to take a photo of their food every time they ate. He was then able to determine the glucose response of each meal, and how long it took for the blood glucose level to come back to baseline.[14] For some people, the same food elicited a sharper rise in blood glucose, while others eating the identical food at the same time had a shallower rise. This type of analysis could be used to determine if someone has a larger glucose rise at night. Then individuals could figure out when they should eat their last meal in order to produce a modest rise in blood glucose. Or, conversely, they may be able to find out if a dinner rich in protein and fat will produce a smaller rise and hence be a healthier choice for them. This is an ideal sensor to use when you are starting with TRE and are worried that your glucose level will drastically decrease at night, as you can track the data on your smartphone. Currently, these devices are not directly marketed to consumers in the United States. They can only be ordered and prescribed by a supervising physician. As these sensors are rapidly evolving, many are moving from medical-grade to consumer-grade, so speak with your physician or local pharmacist about availability.

PART III

Optimizing Circadian Health

CHAPTER 9

The Clock, the Microbiome, and Digestive Concerns

Sandy thinks she is in perfect health, except for the antacids she takes every night before she goes to sleep. Tom is sure that a diet high in gluten is contributing to his daily stomach pains and digestive issues. Lisa knows that she can't tolerate dairy. Abby doesn't know why she is chronically constipated. Maria can't sleep through the night unless she eats a bowl of ice cream before bed.

These digestive issues are so common that many of us don't think they fall under the category of health concern, let alone chronic illness. According to the National Institute of Diabetes and Digestive and Kidney Diseases (a part of the National Institutes of Health), it is safe to say that more than three-quarters of the U.S. population suffers from one or more chronic digestive ailments, including acid reflux, diarrhea, constipation, gas, bloating, and abdominal pain, and most don't report these to their doctor because they dismiss them as normal. However, these symptoms are not normal, and they can be a sign that your digestive system is out of whack. You don't have to live with this discomfort. By tweaking your lifestyle and paying more attention to your circadian code, you can restore your health.

We used to think that the digestive system was like a constantly active boiler, where you could add food at any time and it would get metabolized to create energy. Now we know that this is not the case. Almost every aspect of eating, from craving food or feeling hungry to digestion and elimination, occurs according to strong circadian timing. What's more, we also know that eating the wrong foods at the wrong time not only disrupts the digestive clock but creates disease and chronic illness.

The Rhythms of Digestion

The digestive process is divided into stages, and each stage has a circadian component. The first stage, the cephalic phase, occurs in the mouth. Like Pavlov's dogs, when we see food, think about food, or are accustomed to eating at a certain time, our mouth begins to produce saliva that is rich in enzymes, making it easier for the stomach to do its job. The mouth produces even more saliva as soon as we start chewing, while the brain instructs the stomach to release digestive acids. Nearly one-third of the acid needed for digestion is released in the cephalic stage. Even a small snack after dinner—a piece of chocolate, a glass of wine, even an apple—triggers the secretion of gastric acid, starting the entire digestion process, which then lasts for hours. This disrupts the circadian program: In the evening, when we are supposed to be cooling down, eating new foods warms the body, making it harder to go to sleep.

Saliva secretion is circadian: It is most productive during the day, up to 10 times greater than it is when we sleep. The nighttime drop in saliva production helps us stay asleep, although it is another reason we wake up with dry mouth. Daytime saliva secretion neutralizes stomach acid that may come up through our esophagus into our mouth, but reduced saliva at night is not sufficient to carry out this task. Eating late at night can produce excess stomach acid, and if that acid comes back up the esophagus and into the mouth, there is not enough saliva to neutralize it. As a result, late-night eating can trigger acid reflux, causing inflammation of the esophagus and permanent damage to the esophagus, stomach, and teeth if left unchecked.

Once food is properly chewed and swallowed, it travels down the esophagus and passes into the stomach, beginning the gastric phase of digestion. The acidic environment of the stomach is like a brewing vat, further breaking down food into microscopic particles. The acid is contained in the stomach by the sphincter muscle that is at the junction between the esophagus and the stomach. This acid is so strong that it can even kill bacteria found in raw food like salad or sushi. Excess acid production, even at the right time of the day, causes acid reflux. Diminished acid production is also bad, because it promotes the growth of dangerous bacteria that cause diarrhea. It also allows for incompletely digested food particles, which can trigger inflammation by the immune cells present in the gut lining. This is referred to as a *leaky gut*.

The stomach lining is covered with a cushion of mucuslike material that ensures it does not get damaged when food particles pass through it. This lining is filled with cells that are arranged like a cobblestone street. When any one of those cells gets damaged, the lining becomes compromised, opening the possibility of gut content leaking into the body. Both mechanical and chemical actions during the digestion process damage these cells, and the lining gets repaired between meals. Individual cells, if damaged, can be removed and replaced by new cells. In fact, there is so much damage to our gut lining that 10 to 14 percent of cells are replaced every day. This repair and replenishment process is circadian. Every time we sleep, growth hormone secreted from the brain acts on the gut lining to repair itself, instructing the gut lining to check for damaged cells and replace leaky patches with new cells. The cells also secrete copious amounts of mucus to grease the gut lining, as some mucus gets depleted with every meal.

Stomach acid production and secretion happens every time we eat, and there is a circadian component to that as well. Stomach acid production is typically high during the hours before bedtime, roughly 8:00 to 10:00 p.m.[1] If morning stomach acid is produced at an arbitrary unit of 1, at night it reaches 5. However, when food is consumed during the day, your stomach acid production may go up to 50; eating the same amount at night may increase production up to 100. This means if we eat a modest meal in the evening, the stomach will produce a larger amount of acid than if the food was consumed at noon. This may be a defense mechanism of the gut to make sure that if a bacteria or pathogen were to somehow make its way to the stomach at night, the acidity of the stomach could destroy it before it got to the next phase, the intestinal phase, which slows down at night. Therefore, any food that enters the stomach at night has to wait in a high-acid environment. Excess acid produced in response to a late-night meal fills up the stomach, and as food moves slowly along the digestive tract at night, this acid slowly creeps up and can come up to the mouth, causing acid reflux.

Our food sits in the stomach for 2 to 5 hours, depending on how much we eat. Then it passes from the stomach to the intestines, where further enzymatic and chemical digestion continues. This marks the beginning of the intestinal phase. The intestines are not designed to handle the high acidity that is present in the stomach, so once the food enters the intestines, acid secretion is reduced and neutralized.

Once food enters the intestines, it does not move by itself. Rather, it is squeezed along the digestive tract by muscles that surround the tube. This is called *gut motility* or *gut contractility*. An electrical signal from the gut's nerve cells triggers the muscles to expand and contract. This produces a wavelike motion that pushes food through the tube. Once food is fully digested and the nutrients absorbed, the waste by-product reaches the colon, the last part of the gut, and exits the body as stool, a full 24 to 48 hours later. This movement from intestines to elimination has a circadian rhythm: It is more active during the day, while at night the movement is very slow. This is why we don't typically wake up in the middle of the night to have a bowel movement. Eating a heavy meal and lying down immediately afterward does not allow food to move down the intestinal tract as fast as it should, and this also leads to acid reflux. This becomes more evident as we get older. Just like our muscles weaken as we age when we don't engage them properly, our stomach muscle can also weaken. When this happens, the electrical impulses that push food down the stomach become weak as well, and when we are in a horizontal position, without the force of gravity helping us, the food will not pass through the intestines and will stay in place or move very slowly.

Instead of lying down and watching TV or other screens after dinner, a better habit to adopt is taking a short walk or doing some chores that require standing. Working with gravity, rather than against it, helps prevent reflux.

Is There Really a Leaky Gut?

When the gut leaks like an old garden hose, it can expose internal organs to digestive enzymes and bacteria, causing immediate, life-threatening septic shock. This situation requires immediate medical care. For simplicity, I will use the term *leaky gut* as many other medical experts do: to describe a gut that is not in good condition, is prone to inflammation, and may leak very tiny particles that may be smaller than average bacteria.

What's more, foods you may be sensitive to do not have to leak out of the gut to cause systemic inflammation. If these offenders come in contact with the stomach lining, there are enough immune cells in the gut that become activated and will start the inflammatory response. These same immune cells from the gut can travel to the rest of the body via the bloodstream. When these immune cells are activated, they "spread the word" about the offending foods and spread inflammation. These two explanations show that foods do not have to breach this barrier in order to cause health issues.

All Foods Are Not Digested Equally

Each type of food macronutrient—proteins, carbs, and dietary fats—is digested differently. All nutrients are first absorbed into the stomach lining, which releases them to a special bloodstream that only carries blood from the intestine to the liver. From there, the nutrients go to other organs. Proteins are broken down into amino acids that are easily absorbed into the bloodstream to be used as building blocks for new cells. Carbs are broken down into simple sugars. Dietary fats are the most difficult to absorb. They require bile, produced in the liver and stored in the gallbladder, to convert them into an emulsion, which is later taken up in the small intestine and then in the bloodstream. Production of bile is strongly circadian. This rhythm not only makes sure that sufficient bile is ready to absorb fat from our diet, it also breaks down cholesterol in the liver.

The absorption of glucose, amino acids, and fat is strongly circadian. Nutrient absorption requires a lot of energy, which is why it can't happen all the time. Gut cells that absorb these nutrients and other chemicals in food have different channels or doors that allow only certain types of molecules to go through, and the opening and closing of these doors is circadian.

During digestion, each macronutrient also activates different gut hormones. Amino acids (from proteins) activate the hormone gastrin that instructs stomach cells to release acid. Similarly, fat activates the cholecystokinin (CCK) hormone in the intestine, which in turn releases bile from the gallbladder. Many of the hormones and chemicals produced in the gut stimulate the brain to affect our emotions and cognition. For instance, CCK and other hormones produced in the gut affect whether we feel depressed, excited, anxious, or panicky.

Other gut hormones sense the presence of food and send a signal to the rest of the body and the brain that a new source of energy is available. For example, when the stomach is empty, the hormone ghrelin signals the brain to feel hungry. Ghrelin itself has a circadian rhythm to make sure our hunger aligns with the stomach being empty. After a meal, our ghrelin level goes down and makes us feel full so that we stop eating. If our ghrelin level is not in sync, then we continue to feel hungry even if we've had a full meal. At that point, our stomach would have too much food and not enough digestive juices, which could lead to indigestion. Sleep reduces the production of ghrelin so that there is less chance of us waking up and needing to eat. But when we don't get enough sleep, even if our stomach is still digesting our last meal, our ghrelin level goes up and makes us think we are hungry.

This reaction may be our body's preparedness mechanism for our brain to make sure that we have enough energy for an unexpected emergency at night. Our ancestors did not wake up in the middle of the night to take a phone call or check texts/e-mails. They woke up to run away from a predator or to douse a fire, events that required a lot of physical activity. So, there is a natural emergency program that keeps us eating into the night in order to make sure we'd have enough energy if we had to wake up and move quickly. This may be one reason why short sleep duration is associated with a high ghrelin level and, ultimately, obesity.[2,3] Following a time-restricted eating plan improves sleep and the daily rhythm of hunger and satiety so that you feel less hungry at bedtime.

The Gut-Brain Axis: Anxiety and Circadian Disruption

Sometimes, the CCK hormone is partially broken down into a smaller hormone called CCK-4, which is quite dangerous, especially if it gets into the bloodstream and travels to the brain. When it reaches the brain, it can turn on the brain switch for anxiety, panic attack, and unnecessary fear. This process is so potent that one needs only 1/20 of a milligram of CCK-4 injected into the bloodstream to produce a full-blown panic attack.[4]

Sleep disturbances can increase anyone's predisposition to anxiety, but the underlying mechanism is largely unknown. We believe that a sleep-deprived individual or someone who goes to bed late is more likely to eat late, which triggers CCK production. If there is a defect in CCK breakdown and CCK-4 accumulates in the blood, this might explain the increased incidences of anxiety in sleep-deprived people.

Proving the Existence of the Gut's Clock

As you see, there are so many interrelated processes that occur inside the gut that it is not easy to reset the clock in each of these different parts. That may be the reason why the gut's clock takes the longest to adjust to a new time zone. When you have jet lag or stay up too late at night, your food may take longer to digest, you may get acid reflux, and the next morning your bowel movement may not be right or you may have constipation.

To prove the circadian relationship of gut function, Carolina Escobar, a professor at the National Autonomous University of Mexico, did a simple experiment.[5] She measured the clocks in different organs of rats that had free access to food. Then she changed the light-dark schedule in the rat house as if the rats had traveled across six time zones. For the next few days, she monitored how the clocks in different body parts changed their timing to adjust to the new light-dark cycle. She found that the gut clock was the slowest to reset to the new time zone. In a second experiment, when she changed the light-dark cycle, she also made the rats eat only at the new local time (no free access to food). In this case, the gut's clock took fewer days to adjust to the new time zone, and the rats were less prone to the discomfort of jet lag. This is one way we know that maintaining a regular eating-fasting cycle ensures the gut's internal timing is in sync with your determined eating schedule. Similarly, one of the tricks to beat jet lag is to make sure that if you are up at night in the new time zone, you resist your temptation to eat until morning. Eating at the time appropriate to the new time zone is the best way to reset the gut clock.

Gut Function Affects Overall Health

If good nutrition is the key to optimally fueling our body, then the gut is the gateway through which nutrition gets into our system. Most diseases of the gut compromise its core function of absorption of all nutrients, minerals, and vitamins for our body. For example, when people have an allergy to the gluten protein, found in wheat, barley, and rye, eating wheat products produces an inflammatory response in the gut. If untreated, this can cause uncomfortable long-term digestive problems. And, when our digestion is off, we don't feel right, which can affect our sleep, our productivity, and our drive to exercise.

What's more, if the gut function to absorb any specific nutrient or mineral is compromised, then the rest of the body suffers. When the rest of the body cannot get all the nutrients it needs, it can develop diseases, such as anemia from low protein absorption or fractures caused by insufficient calcium.

Simon's Anxiety Started in His Gut

Through our myCircadianClock app and from personal feedback from people who have adopted a TRE lifestyle, we have learned about the potential connection between eating patterns and anxiety. For instance, Simon was overweight—his doctor told him he needed to lose 30 pounds—and he experienced occasional panic attacks. One of the things he was particularly worried about was his general health. He adopted a 10-hour TRE to see if he could lose weight and increase muscle mass.

Simon's diet was actually very clean even before he tried TRE. He kept track of what he ate, and we could see that he ate a balanced diet, counted calories, and went to the gym regularly. So, there was actually very little room for improvement when it came to the kinds of foods he ate: He just had to focus on when he would eat them. When we told him this, Simon didn't take it as good news. It only made him more anxious that he was eating the right things but still gaining weight.

After a few weeks of following a 10-hour TRE, Simon noticed a marked reduction in his general anxiety and panic attacks. He also noticed that he was sleeping better. This general reduction in anxiety actually helped him focus on his tasks, including staying on the 10-hour schedule.

Simon also reported that he was losing weight steadily, 1 to 2 pounds a week. I am not sure if it was the improvement in his sleep or the reduction in his waistline, or how either of these affected the other, but our research did show that the gut-to-brain signal for relieving general anxiety was activated. We also know that when you can stay calm, you are more likely to follow through on your tasks. Lowering Simon's anxiety was an important step to get him to focus on his TRE and lose the weight.

The Gut Microbiome Is Circadian

Every part of the digestive tract is filled with microbes or bacteria, each requiring a different environment to grow and thrive. Some bacteria like it more acidic, some more neutral; some like to feed on protein, some on fat or sugar; and each maintains its own eating-fasting rhythm. There are some species of microbes that flourish under fasting, while others flourish during feeding. Therefore, the composition of the gut microbiome changes between day and night. In other words, we go to bed at night with one set of bacteria in our stomach and wake up with a different set, and in the middle of the day a different set appears.[6] Each type of bacteria has a different function and digests different types of nutrients. For instance, many food components cannot be broken down by gut enzymes, and those food components require the gut microbiome. Dietary fiber and other chemicals present in food can only be digested by gut microbes that live in the intestines. Therefore, maintaining a diverse mix of gut microbes is considered key to a healthy gut.

One way to maintain a diverse gut microbiome is to eat a diet that has diverse sources of nutrition. Researchers found that when mice eat a high-fat/high-carb diet randomly throughout the day and night, their gut does not receive a rich diversity of foods to support all of the necessary bacteria.[7] When their gut microbiome breaks down and only a few varieties of bacteria are left, the result is obesity. We believe the same is true in humans: Without all the right bacteria we cannot fully digest our food, and the rest is stored as fat.

We also know that when we experience poor sleep or conditions that simulate jet lag or shift work, the gut microbiome composition is altered to a state that supports obesity.[8] For example, when fecal matter from jet-lagged people is placed in the gut of healthy mice, the mice become obese. But fecal matter from healthy people, who have not traveled or are not doing shift work, does not trigger obesity in rodents. These observations have raised a lot of interest in understanding how shift work, jet lag, and circadian disruption change gut microbes so severely that they can in turn nudge the body toward obesity.

You may think that when we travel, airport food does not provide the healthiest options, so it may be that bad food promotes bad bacteria, which then causes obesity. If this is true, then we won't ever escape from the bad food–bad bacteria spiral. However, we did a simple experiment with mice:

We gave them a high-fat/high-carb diet but kept them to a strict feed-ing-fasting cycle, and the mice remained healthy.[9] Under TRE, one set of bacteria flourishes when mice eat, while a different set populates the gut during fasting. Overall, a mixture of good gut bacteria flourished under TRE, while several harmful species that promote obesity or disease were suppressed. This research is very encouraging: If these findings hold true for humans, then shift workers who may only have access to unhealthy foods can still maintain a healthy population of microbes in their gut and prevent obesity, and the diseases associated with it, if they keep to a good TRE cycle.

We found that TRE in rodents optimizes the gut microbiome in such a way that the gut most efficiently processes and absorbs nutrients and excretes waste, and this produces better health. The gut microbiome under TRE changes the breakdown and absorption of fiber in a way that a good chunk of the sugars are not absorbed but leave the body during elimination. TRE also changes the gut microbiome in a way that con-verts bile acids into a different form that is excreted in stool. Since bile acids are produced from cholesterol, the more bile acid that leaves the body, the greater the reduction in cholesterol in the blood.

The Gut Microbiome Affects Our Food-Mood Axis

The food we eat and microbes in our gut work together to produce sev-eral hormones and chemicals that impact our mood and can determine whether we feel calm, anxious, depressed, manic, or panicky. The right amounts of gut bacteria convert some of our food into the neurotrans-mitters that keep our brain balanced and working effectively, including dopamine, gamma-aminobutyric acid (GABA), histamine, and acetyl-choline. However, some bacteria in the gut cause some carbohydrates to ferment and produce fatlike molecules called *short-chain fatty acids* (SCFAs), which then have negative effects on our health. SCFAs can travel to the brain and affect brain development and function.[10]

Gut bacteria also influences the effectiveness of certain medications and produces chemicals that act like drugs. For example, many antibiotics can change the composition of our gut microbes, and at the same time the surviving microbiome can convert antibiotics into chemicals that affect brain function. This may explain the side effects of some antibiotics, such as anxiety, panic, depression, psychosis, and even delirium. In babies and toddlers, the unintended effect of diet and medications can have a lifelong impact. For instance, the gut microbiome is now increasingly recognized as a contributing factor in autism. [11,12]

Choose Foods That Protect the Microbiome

Food preservatives have a very detrimental effect on the gut. Have you noticed how the food you cook in your own kitchen cannot stay fresh for more than a few days in your refrigerator, while the packaged food you buy at the supermarket will not spoil for a long time? Preservatives are added to food to inhibit the growth of bacteria that spoil food. When these preservatives get into our intestines, even at a low concentration, they inhibit the growth of gut bacteria, affecting the composition of the gut microbiome.

Some food preservatives, such as carboxymethylcellulose and polysorbate 80 (an emulsifier used to make foods like ice cream smoother and easier to handle, as well as to increase resistance to melting), also have detergent-like properties that inhibit bacterial growth by thinning the protective coating around bacteria cells. However, our gut mucosa has a similar coating. Food preservatives can corrode the protective mucosal lining that separates microbes from the cells that line the gut. When these unwanted microbes make contact with the cells in the gut lining, it can cause inflammation, such as colitis.[13,14] TRE promotes repair of the gut lining and may counteract the negative effects of a bad diet.

An assortment of different types of food that includes lots of different fresh fruits and vegetables promotes the healthiest gut microbiome. The good bacteria in the gut feed on the dietary fiber found in fruits, vegetables, and complex carbohydrates. When we don't eat enough fiber, it's like eating food with lots of preservatives: The microbes in our gut that have nothing else to eat will instead dine on the gut's mucosal lining.[15]

Circadian Disruptions Cause Diseases of Digestion

When food arrives regularly at the same time, all the clock processes in the digestive system work together for efficient digestion and elimination, and the gut remains healthy. When food arrives at a time when the gut is not anticipating it, like in the middle of the night, that food may not be digested properly and can also interfere with the normal repair process of the gut and leave physical damage. Over time, such damages accumulate and can lead to diseases of the gut.

If we eat three meals every day, say at 8:00 a.m., 1:00 p.m., and 6:00 p.m., our gut learns to anticipate these meals, and it floods the gut with digestive enzymes and acids only after we start eating. If we miss a meal, not much damage happens. But when we eat in the middle of the night when the gut is being repaired and there is not much gut contractility, more damage is done.

Just one day of eating late at night may leave you with a bad feeling in your stomach the next morning. If it continues for a few days, acid reflux may increase, and your gut may not have enough time to repair all the damaged cells in the gut wall.

If random eating continues for many weeks, acid reflux and heartburn (known as *gastroesophageal reflux disease,* or GERD) may become a permanent fixture in your life. Indigestion, irregular bowel movements, or constipation may become a part of your daily struggle. Normal gut bacteria composition will change, causing leaky gut. This can cause both local inflammation in the gut and bodywide inflammation; those symptoms include general fatigue, joint pain, skin rashes, arthritis, and food sensitivities. As the immune system fights this unnecessary battle, it weakens when it has to fight real pathogens. You may be more prone to bacterial infections you would otherwise be able to address. These ailments might exacerbate to Barrett's esophagus, esophagitis (inflammation of the esophagus), tooth decay, peptic ulcers, inflammatory bowel disease, and even colon cancer.

We know that a circadian disruption is at the heart of many of these issues because shift workers are predisposed to diseases of the gut. In fact, in a study of more than 10,000 shift workers, researchers found shift work doubles the chance of developing gastric ulcers and duodenal ulcers.[16] And since we are all shift workers, it makes perfect sense that nearly 10 to 20 percent of the population in the developed world experiences acid reflux at least once a week, and in the United States alone, more than 60 million prescriptions are written every year for GERD.

Taking Acid Medication for Months Is a Bad Idea

So, what is the big deal about having GERD or acid reflux? You pop a pill and the symptom disappears—almost like taking a mint for bad breath. Nope. A survey conducted by the Gallup organization on behalf of the American Gastroenterological Association found that of 1,000 adults experiencing acid reflux at least once a week, 79 percent reported experiencing heartburn at night. Among those, 75 percent reported that symptoms affected their sleep, 63 percent believed that heartburn negatively affected their ability to sleep well, and 40 percent believed that nocturnal heartburn impaired their ability to function the following day.[17] Clearly, GERD was affecting their circadian rhythms.

However, medication wasn't helping. Of the 791 respondents with nighttime heartburn, 71 percent reported taking over-the-counter medicine for it, but only 29 percent of those respondents rated this approach extremely effective. Forty-one percent reported trying prescription medicines, and close to half of those respondents (49 percent) rated this approach extremely effective. This means that the expected result from medications for heartburn was not achieved by a sizable percentage of patients. So why do people keep taking them?

Most antacid medications essentially slow down acid production in the stomach. But this is only a temporary fix, and like overused sleep medications, antacid medications have never been tested for continuous use for months or years. The drugs that belong to the class are called *proton pump inhibitors* (PPIs); having more protons in the stomach means more acidity, so PPIs essentially inhibit molecules that pump more protons into the stomach. As you can imagine, these drugs change the pH of the stomach and make it less acidic. But the body fights back and tries to make more acids or more of the hormone gastrin, which tells the stomach to make more acids. This may lead to dose escalation. Once one uses PPIs regularly for a few weeks or months, the gut chemistry also changes in such a way that the person becomes dependent—even addicted—to PPIs.

As stomach acid reduces, more bacteria can survive in the stomach and enter the small intestine—some of which can be pathogenic. This is how PPIs can lead to infections and diarrhea. A systematic review of six different studies on more than 11,000 patients using these medications showed a threefold increase in salmonella infection.[18] Similarly, a second longitudinal study on more than 14,000 middle-aged adults who used PPIs found an average of a threefold increase in bacterial infections of the stomach.[19] Some of the participants were even more susceptible: Their risk was as high as tenfold.

PPIs also increase the risk of kidney diseases. In studies involving more than 500,000 patients in New Zealand and 200,000 patients in the United States, regular PPI use was found to increase the chance of acute kidney disease or acute inflammation of the kidney by threefold.[20,21] The adverse effects of PPIs extend even to the brain. There are some studies that show chronic PPI users may have an increased risk for dementia. PPIs are also used for the prevention of a host of other diseases, including stress ulcers, peptic ulcer disease, gastrointestinal bleeding, and *H. pylori*.[22]

Continuous use of these drugs is also linked to changes in bone density, causing osteoporosis and bone fractures.[23] Medications for these diseases are known to affect gut function, including causing constipation. This is how "drug use begets drug use" and we get into a spiral of using another drug to manage the adverse side effects of a previous drug.

This spiral can be stopped or slowed down with some simple changes to our lifestyle, including when we eat and when we go to bed. A combination of TRE, exercise, and sleep will promote optimal digestion, reduce intestinal permeability, and improve overall gut health. Improvement in your gut health may help you wean off or reduce the amount of medications you take for these gut diseases. Reduced medication may further benefit you by reducing adverse side effects.

We find a majority of TRE practitioners say their GERD reduces once they establish and follow an eating schedule. It is such a common benefit that some people don't even mention it and focus instead on how TRE benefits something worse about their health. However, the more we can highlight digestive issues like GERD, the more we will all realize that it is not a normal part of living, or something we just have to learn to live with.

Eating Patterns and Irritable Bowel Syndrome

Irritable bowel syndrome (IBS) is a type of gastrointestinal disorder. IBS symptoms and signs include:

- Abdominal pain

- Altered bowel habits (more or less)

- Bloating

- Cramping

- Gas

We have recently noted that mice fed a standard diet eat small snacks here and there, and their bowel movements remain highly cyclical. When the same mice were fed highly processed food and allowed to eat all the time, they also pooped all the time, as if they had IBS. But restricting their diet to a few hours completely took care of their frequent pooping and restored the daily bowel movement cycle. This raises some hope that people with IBS may benefit from TRE.

IBS is rapidly rising among teens and young adults. Although there has not been much study on what is driving the increasing incidence of IBS in young people, one hypothesis is that sleep and circadian rhythm disruption begins during middle school and high school, when students begin to stay awake late into the night, have late-night snacks, and sleep less. Circadian rhythm disruption among teenagers may be the trigger for increasing incidences of IBS.

Some practitioners of TRE have reported they do see improvement in their IBS symptoms after just a few weeks. For instance, Patty was in her early forties and had been suffering from IBS for more than 7 years, with at least half a dozen trips to the bathroom each day. She started an 8-hour TRE, with her first meal at 10:00 a.m. and the last meal at 6:00 p.m. After 2 weeks, she e-mailed us, reporting that she had improvement in her IBS symptoms that had not been possible using any medication.

My hope is that anyone with digestive complaints who tries TRE for at least 12 weeks will have the same experience as Patty. It continues to amaze me that by making one small change—adapting a new eating pattern that is naturally aligned with how our body is supposed to work—can quickly deliver better health.

CHAPTER 10

The Circadian Code for Addressing Metabolic Syndrome: Obesity, Diabetes, and Heart Disease

Dear Dr. Panda,

Just to inform you, as of yesterday I officially lost 40 pounds. I have been on an 8-hour TRE for 3 months, and as of August 1 I lost a total of 40 pounds. I have learned so very much about my body and what it truly needs to function and flourish. My next goal will be achieved when I lose another 10 pounds.

I started the diet at 300 pounds and now weigh in at 260. My life has totally changed, and I feel like I am in control of my body and my relationship with food has changed forever. I lost 10 pounds the first week and then nearly plateaued for 2 to 3 weeks before the weight started melting off. I found that if I eat less "heavy" or harder-to-digest foods earlier in the day that I will have a better shot at burning more stored fat before breaking the fast the next day. I believe that this is the answer that so many people have been searching for. I have about 20 friends that are doing the diet with me and have achieved great success. I spoke with a truck driver yesterday who said he had essentially given up and did not think there was a way out at 400 pounds. I explained how the diet works and gave him hope.

Bottom line is that your diet works. I am a huge advocate for it and will continue to help as many people as I can to take control of their lives and no longer be victims of their poor eating habits.

Weston "West" Barnes

Metabolism is the chemical reactions that occur in the body to use the nutrients we eat to produce energy, make the building blocks to repair and grow cells, and eliminate waste. When our body's metabolism goes awry, it throws off the digestion of fat, sugar, and cholesterol, and we gain weight. These added pounds affect our health in the form of metabolic diseases: obesity, diabetes, and heart disease. This trifecta can happen together or separately, but when you have symptoms of one, symptoms of the others can slowly appear. As these diseases and their symptoms accumulate, they affect the normal function of the rest of the body. This is referred to as *metabolic syndrome.*

Your doctor uses simple criteria to test if you are on the path to metabolic syndrome. *The Third Report of the National Cholesterol Education Program (NCEP) Expert Panel on Detection, Evaluation, and Treatment of High Blood Cholesterol in Adults (Adult Treatment Panel III)* defines metabolic syndrome as the presence of any three of five traits:

- Abdominal obesity

- High blood pressure

- Laboratory abnormalities of triglycerides (a type of fat in the blood)

- High-density lipoprotein-cholesterol (HDL-C) levels

- Fasting hyperglycemia (a signature of diabetes)

Metabolic syndrome can be lethal, but it is also completely and totally preventable and reversible. Weight loss, exercise, and adapting to a healthier circadian code are the keys to preventing and reversing this disease. The key is to lose body fat, and particularly abdominal fat. Abdominal fat actively produces harmful pro-inflammatory molecules and other chemicals that cause atherosclerosis and cancer, elevate blood sugars and insulin resistance, and contribute to inflammation. By following a TRE plan and combining it with vigorous exercise, you have the best chance of losing inches around your waistline and reversing your health.

A Break in the Circadian Code Can Lead to Obesity

As soon as we eat something, our pancreas releases insulin, which does two important jobs for metabolism: It helps absorb sugar from our blood into our liver, muscle, fat, and other tissues, and it signals these organs to convert some of the sugar to body fat. This process continues for up to 2 to 3 hours after we eat, every time we eat. So, as we keep on snacking, our body remains in fat-making mode. In the first half of the day, the pancreas produces more insulin, and at night it slows down. The body remains in fat-making mode for a longer time after a late-night meal. Only after 6 to 7 hours of not eating does our body begin to start burning some fat. This is the critically important aspect of TRE: to stop feeding the engine that is your body and let it run on the fuel it already has. This is the only way to prevent or reverse weight gain and, ultimately, obesity.

Obesity is generally described as excessive body weight relative to height. The traditional and most widely used definition of being obese relates to body mass index (BMI). The American Medical Association defines obesity as having a BMI of 30 or higher. Obesity is more than just being overweight; it can affect the rest of your health. It puts you at a greater risk for developing fatty liver disease, diabetes, hypertension, heart disease, and chronic kidney disease. These diseases are related to where you store extra body fat.

Excess energy beyond what can be stored as glycogen is converted to fat and is stored as fat in our adipose tissue, or fat cells. When the adipose cells reach their full capacity, our body tends to store fat in cells or organs that are not designed to store it. This compromises the function of organs such as the liver, muscles, and pancreas. When there is excess fat in cells, there is less space for the cells to carry out their normal tasks of generating energy. This factor is linked to a range of diseases from fatty liver disease to diabetes, heart disease, high blood pressure, and even cancer.[1]

When cells carry excess body fat, there is also less space for the endoplasmic reticulum (ER), the canal system within a cell that connects to the cell membrane and then to the outside of the cell. Cells always secrete something through this canal during the daily repair cycle. But when the ER is stressed, the cell's overall repair process is hampered. Some body fat is also converted to the type of fat that causes inflammation and is released into the blood. These inflammatory fats can contribute to inflammation all over the body.

Disrupting your circadian code is a major contributor to obesity. First, reduced sleep confuses the brain hormones that regulate hunger. The brain cannot predict how long a person is going to stay awake, and since staying awake requires more energy than sleeping, the brain increases hunger hormone production. As a result, people always eat more than what is needed to stay awake for just a few extra hours. Sleep deprivation confuses the brain, making us choose unhealthy foods over healthier options. We crave energy-dense foods when we are overtired, and overeating these foods ultimately contributes to obesity. Sleep deprivation also makes us lethargic and less active, which further contributes to excessive energy storage.

Every time we eat, our pancreas produces insulin to help the liver and muscles absorb blood sugar. At the same time, insulin promotes the biochemical pathway that makes fat from sugar. When we spread our calorie intake over a long period of time, it keeps insulin production active, which tells our organs to keep making body fat. Shifting meals later in the evening or into late night when we are physically less active further contributes to reduced energy expenditure and more fat storage. Lastly, increasing the number of hours we eat makes it so the body is never allowed to burn off stored fat, as it is constantly using newly digested foods for energy.

TRE Creates New Eating Patterns

The old adage for improving health has been to eat many small, nutritious meals spread throughout the day.[2] Even my own personal trainer recommended eating every 2 to 4 hours up until my bedtime. This eating regimen was devised for two extreme ends of the population. Physicians thought that people with prediabetes should eat small meals to reduce the flood of sugar that rushes through their arteries after each meal, so that smaller amounts of insulin produced in their pancreas could handle the blood glucose rush. The other population was athletes who were training for bodybuilding events or triathlons. They believed eating frequent meals was a good strategy for keeping the body in an anabolic, muscle-building mode so that they could build more muscle. In reality, the outcomes of eating this way are mixed, and it is not recommended as a lifelong habit for anyone, regardless of their level of exercise.

The general population does not belong to either of these extremes. While people with diabetes need to eat small meals to keep their glucose levels from spiking, for the average person it's hard to keep calorie intake low if we are constantly eating, even if they are small meals. Additionally, even for patients with prediabetes, the recommendation to eat small meals does not imply that they should keep eating during their entire waking period of 16 to 18 hours. TRE is a better method of eating because you are training your body to adapt to your natural circadian code rather than an artificial schedule.

This small-meals eating regimen introduced the idea of "healthy snacks" over the past 40 years. Based on National Health and Nutrition Examination Survey (NHANES) data from 1971 to 2010, consumption of snacks as a proportion of total calories has increased from one-tenth to one-quarter.[3] Along with more snacking, total calorie consumption has also increased.

When we reviewed eating patterns from the myCircadianClock app, we found that the traditional breakfast-lunch-dinner pattern is no longer observed, even among healthy non–shift-working adults. In fact, the number of eating occasions ranged from 4.2 times a day to 10.5 times a day. This study clearly demonstrated that 50 percent of adults in the United States eat for 15 hours or longer.[4] This eating pattern may not be unique to the United States, as a similar eating pattern was also found in a study of adults in India.[5]

Hack Your Code: Get Help for Night Eating Syndrome

If you have trouble controlling your after-dinner eating or if you wake up in the middle of the night to eat, you may be suffering from a rare medical condition called night eating syndrome (NES).[6] It is generally believed that NES may result from depression, anxiety, stress, or poor results from attempts to lose weight. As food consumed at night is usually made of high-glycemic carbohydrates, people with NES may suffer from being overweight.[7]

In collaboration with Ying Xu, a professor at Soochow University in China, we have studied mice with NES, and we believe that there may be a genetic component. Some of our mice have a mutation in their Period 1 gene that can cause night-eating-like behavior. These mice start eating in the early afternoon and put on more weight than those who eat at their normal time. However, when these same mutant mice are allowed to eat only at night (when they are supposed to eat), their weight gain slows down.[8] This is a remarkable study, because it shows that if a genetic mutation predisposed the mice to being overweight, imposing a TRE can counteract the bad effect of a genetic condition and keep the mice slimmer.

No such Period 1 mutation has been found in humans, yet. But in the coming years, we may learn more about our own genetic mutations and eating patterns. Until then, one strategy to manage night eating syndrome is to be aware of it and adopt TRE, which will help fight the urge for late-night eating. If it is impossible to hold off your urge to eat late at night, you can try a late TRE, in which you start eating your first meal around lunchtime, so that the last of your food for the day is consumed around midnight. This may not be the best approach to control NES, but it might lessen the overall effect of weight gain.

Alexander Had Night Eating Syndrome

Alexander was 5 feet 9 inches tall and weighed 265 pounds when he contacted our lab. He had gained more than 80 pounds since 2013. He had been night eating for more than 20 years before he contacted us and tried TRE. He told us that he ate in his sleep and that he did not remember what he ate the following morning. Initially, he thought it was because he would deny himself carbohydrates during the day. He had been following a "bodybuilder" lifestyle for more than 15 years, eating a diet that was very high in protein. But the older he got, the harder it became to control his diet, and the binge night eating began.

Alexander had been to doctors, dietitians, and psychiatrists to no avail: He couldn't stop eating at night. He tried zopiclone to treat his insomnia, but it didn't help. He even did a sleep study, which determined that he had sleep apnea, and afterward he used a CPAP machine to help him breathe more regularly at night.

We worked with Alexander and suggested that he try TRE, but we told him that he could pick the hours he chose to eat. Now he wakes up between 7:00 and 8:00 a.m. He only drinks black coffee and water throughout the day. When he gets home from work around 6:00 p.m., he eats his first meal, which is a healthy combination of protein and fats, and lots of salad and vegetables. He goes to bed at 10:00 or 11:00 p.m. and purposely eats before he goes to bed. Most of his calories are consumed between 6:00 p.m. and midnight. Although such late-night eating is not ideal, this is the best TRE he could do, taking his compulsive night eating syndrome into account.

After one month of practicing TRE and trying to lower his stress levels, Alexander reported that the combination had already proved to be so effective that he called it a "life changer." He told us that his focus had returned and he had lost 10 pounds. Despite the fact that he did not eat at all during the day, he still had plenty of energy.

Poor Circadian Rhythms Are Linked to Type II Diabetes

Diabetes occurs when the pancreas cannot produce enough insulin, or when the cells in the body no longer respond to insulin and absorb glucose from the blood. It can develop with increased consumption of sugary foods, reduced exercise, or obesity. However, there is now mounting data that circadian rhythm disruption can lead to diabetes. For example, a week of reduced sleep can raise someone's blood glucose to prediabetic levels.

As diabetes changes the basic property of blood, complications from this disease can affect the entire body and brain. Chronic diabetes can progress to cardiovascular diseases, foot ulcers, damage to the eye, and chronic kidney disease.

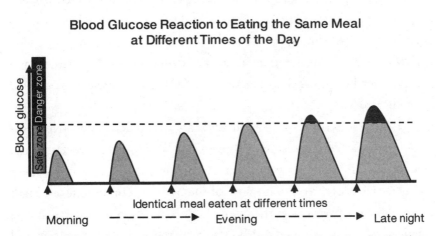

Blood Glucose Reaction to Eating the Same Meal at Different Times of the Day

In the morning, blood glucose levels stay within the safe zone. As the day progresses, the same meal causes your blood glucose levels to rise higher and stay high for a longer period of time.

At least two different circadian clocks are responsible for controlling our glucose regulation mechanism to maintain a daily rhythm. The first circadian clock is in the pancreas, which programs insulin release to slow down to a drip at night. A second circadian clock in our brain produces higher levels of melatonin at night, which act on the pancreas to further suppress insulin release at night.[9] Therefore, if we keep eating late into the night, when our pancreas is asleep, the low insulin drip is insufficient to instruct the liver and muscles to bring additional glucose inside their cells. This leaves the blood glucose levels dangerously high, causing further damage.

Circadian Rhythm Affects Heart Diseases

Heart diseases are caused by obstructions in blood flow. The vast majority of heart diseases are due to deposits of fat in the arterial walls. When blood flow to the heart is compromised, it causes chest pain (angina) or a heart attack (resulting from blood flow to part of the heart being blocked, aka coronary artery disease). When blood flow to the brain is blocked, it is called a stroke or cerebrovascular disease; and when blood flow to a peripheral organ such as the leg is blocked, it is called peripheral artery disease. Another type of heart disease causes the heart to beat erratically. This is known as arrhythmia or atrial fibrillation (also called AFib or AF). AFib is a quivering or irregular heartbeat (arrhythmia) that can lead to blood clots, stroke, heart failure, and other heart-related complications.

Two major causes of heart diseases are dyslipidemia and hypertension, or high blood pressure. Obesity can lead to excessive fat in the blood, which causes inflammation. As the arteries narrow, the flow of oxygen-rich blood to various parts of the body and brain reduces. High blood pressure exacerbates the disease. High blood pressure can dislodge cholesterol plaques, which can flow to a narrower part of the artery and clog it, cutting off blood supply to the brain (stroke), or heart (heart attack).

Disruptions in circadian rhythms affect fat and cholesterol metabolism, leading to increased fat storage, increased cholesterol plaque, and increased risk for inflammation—the recipe for the plaques to form. The circadian rhythm in kidney function produces a daily rhythm in blood pressure, with reduced pressure during the night. This helps lessen the risk of heart diseases. Circadian disruption may keep blood pressure elevated throughout the day and night and can increase the risk for a stroke or heart attack.

TRE to the Rescue!

Time-restricted eating has many benefits for managing metabolic diseases. TRE helps reduce weight, improve glucose control, and maintain hearth health. With all three of these benefits in place, you can see real reversals in disease. Here's how it works:

The most obvious is the fact that TRE reduces your opportunity to eat. Just by consolidating your meals—at first into a 12-hour interval—you naturally reduce your caloric intake. As we discussed in Chapter 5, many really bad food choices are made after dinner, specifically choosing high-fat/high-sugar snacks and alcoholic beverages. If you adopt a schedule in which your last meal happens around 6:00 or 7:00 p.m., there is a good chance that you will cut down on alcohol and all the food that goes with alcohol late at night. And when those snacks go away, you'll automatically be setting your body up for a better, more circadian digestive process and, ultimately, better sleep. The better your sleep, the more accurate your production of the hunger hormones, further reducing your craving for food. And, you're more likely to exercise if you wake up rested, and when you exercise, your brain receives signals to reduce hunger.

TRE positively affects your ability to make good food choices, which ultimately leads to choosing nutrient-rich foods over calorie-dense foods. Starting with breakfast, if you maintain a 12-hour TRE or more, you may find that nutritious foods taste better. You may think this is due to your hunger, which may be partly true. But your taste buds and sense of smell are now better sensitized, which means they are now highly activated. This has an interesting effect on food choice. Slowly, after several weeks of TRE, many people report that calorie-dense foods with lots of added sugar and no natural flavors taste too bland and too sweet: They are no longer attracted to them. Because of this, they don't want to eat as many sweet treats as they used to crave. This magical change automatically makes for better, healthier choices.

Once your body has a longer time to tap into stored glycogen by fasting or exercise, your muscles and liver cells may use up a good portion of glycogen and will make that space available the next day for storing glycogen. After a long period without food, when you eat your next meal, some of the excess carbohydrates will be first stored as glycogen, and there will be less pressure to store them as fat.

Having a clearly defined eating period also revives your hormone production and puts it back in its natural synchrony. The hunger hormone glucagon and its action on the liver is supposed to be limited to a few hours when glycogen storage is depleted. This hormone instructs the liver to make glucose from amino acids when we fast. But if you have obesity and/or diabetes, this program is on 24-7, so your liver continues to make sugar from amino acids even after a meal—causing blood sugar to rise and supplying less amino acids to build muscle proteins. Under TRE, glucagon function can normalize, so the liver reduces its own glucose production by half, sparing proteins to be used for maintaining healthy muscle. This may help in reducing blood sugar.

TRE not only reduces the pressure to store more fat, it also restores your body's own rhythm to burn fat. Your liver and muscle cells need several hours of fasting at night to turn on their fat-burning mechanisms. TRE restarts this process if it has fallen out of rhythm. While a healthy fat cell can devote 90-plus percent of its volume to storing fat, a liver cell with more than 20 percent of its volume as fat is a sick cell. Therefore, even a small decrease in fat in the liver cell has a huge beneficial impact by improving liver function. As the fat store depletes from the liver and muscles over the first few weeks of TRE, it makes room for more storage of glycogen. And as space inside all of your cells becomes more available, the cells become healthier.

We found another interesting connection between cholesterol and fat. TRE increases the level of an enzyme that breaks down cholesterol in the liver. Cholesterol is usually broken down into bile acids. TRE mice showed a reduction in their blood cholesterol to normal levels and a slight increase in bile acids. A small increase in bile acids is considered good as it triggers a program in fat cells to burn off fat.[10]

We also know that systemic inflammation subsides with TRE.[11] Systemic inflammation is the mother of many metabolic diseases: diabetes, fatty liver disease, atherosclerosis, and others. Weight loss leads to less of the inflammatory fat that typically activates the immune cells causing inflammation. With less inflammation, joint pain and soreness decrease, making it even easier to increase physical activity.

Overall, TRE reduces the drive to make and store excess fat, improves fat burning, normalizes cholesterol level, and reduces inflammation. Less fat, less cholesterol, and less inflammation mean there is less of a chance of atherosclerosis or clogged arteries.[12]

After several weeks on TRE, the circadian rhythm of the autonomic nervous system comes back, too. This system controls many functions, including blood pressure regulation. My colleague Julie Wei-Shatzel has patients who have experienced a significant drop in blood pressure—the same that is seen when they start using medications—just by following a 10-hour TRE. Some patients who have very high blood pressure and are on medication have also tried TRE, and they see even more improvement in normalizing their blood pressure. In another independent clinical study led by Pam Taub, a cardiologist at the University of California, San Diego, overweight patients at high risk for heart diseases who have been practicing a 10-hour TRE are seeing significant weight loss and reduction in fat mass.

TRE Makes Metabolic Syndrome Medications More Efficient

Most drugs for metabolic diseases are designed to find key regulators of metabolism and act on them. For instance, the most widely used drug to fight diabetes is metformin, which works by activating a protein called AMP-activated protein kinase (AMPK), which triggers better control of glucose and fat metabolism. Interestingly, TRE mimics the effect of metformin by increasing fat burning during the fasting stage.

Many cholesterol-lowering drugs, known as statins, act on the enzyme that mediates the first step in making cholesterol. The same control point is also clock regulated. Under TRE, the rhythm in this enzyme improves; it naturally turns off for half of the day, which essentially mimics how statins work. Statins have adverse side effects, including muscle weakness and muscle pain. Edie, a patient who had been on statins for many years and always had muscle pain, almost completely got rid of her muscle pain after adopting a 10-hour TRE, and she had an easier time with her medication.

Scheduling Heart Surgery? Pay Attention to Your Code

Time of day plays an important role in how well various medical treatments, from taking medications to surgery, work. In a study of 596 patients who underwent either morning or afternoon aortic valve replacement, during the 500 days following surgery, the incidence of major adverse cardiac events was lower for those people in the afternoon surgery group than in the morning group.[13] The differences in circadian rhythms in gene expression over the course of a day may cause a person's heart to heal more quickly in the afternoon than in the morning. These first few hours of healing strongly determine the outcome and long-term recovery, which is why you want your healing to coincide with your circadian rhythm.

The bottom line is that TRE is not just about weight loss: It's a method for addressing real health concerns. It's also true that addressing your weight is one of the best ways to start to reverse disease. If obesity, heart disease, or diabetes run in your family and you have success with TRE, spread the love! Anyone at any age can benefit from living in sync with their circadian code.

CHAPTER 11

Enhancing the Immune System and Treating Cancer

Just like a well-armed defense system uses different approaches and weapons to address different situations, our immune system is a highly sophisticated arsenal that continuously surveys our body, seeking out foreign agents—viruses, allergens, and pollutants to name a few—or tissue damage. If something is amiss, it deploys the right type of molecules in the appropriate quantity to repair damage or neutralize an attack. Once the threat is taken care of, the immune system retreats from its fighting/deployment mode and goes back to its surveillance duties.

Diseases, infections, and allergic reactions often occur when the immune system is too weak or too aggressive, when it mistakenly launches an attack when there was no foreign agent to begin with, or when it continues to deploy long after the threat is neutralized. When the immune system is ineffective, it can start a cascade of reactions that ultimately create systemic or chronic inflammation.

Diseases and symptoms of a nonoptimal immune system cover a wide swath, ranging from acne, aches, and joint pain to flu, asthma, liver diseases, cardiovascular diseases, colitis, rhinitis (or any other disease that ends with -*itis*), and multiple sclerosis. Over time, chronic inflammation can damage the DNA of our cells, which can ultimately lead to cancer.

For example, people with ulcerative colitis and Crohn's disease have an increased risk of colon cancer.[1]

However, just like the major organs, the immune system has a circadian component, and if you can resync it, you can regulate its response. What's more, disrupting your circadian code affects your immune system, making you more susceptible to diseases or infections and making it more difficult to recover quickly. For instance, wound healing has a strong circadian component. Both bleeding and clotting time have to be exquisitely balanced: You don't want to clot too quickly. A blood clot is like a cement patch on a leak. The bonding structure is made of proteins produced in the liver, which we know is strongly circadian. If we bleed for too long before the clot forms, we can develop an infection.

Shift workers have been shown to have fragile immune systems. Compared with non–shift workers, shift workers have a higher incidence of the inflammatory diseases of the gut (colitis), as well as higher risks for developing bacterial infections, several types of cancers, and many other immune system–related chronic diseases, including cardiovascular disease and arthritis. And if we are all shift workers, then these diseases may be in store for us. In this chapter, you'll learn exactly how your circadian code affects your immune system and how medications, surgeries, and treatments can be better aligned with this code for optimal health and recovery.

The Circadian Clock Controls the Immune Response at the Cellular Level

There are many different types of immune cells in our blood, and they serve different purposes. Each type of cell is part of a distinct immune system. Some destroy bacteria, others repair wounds, and still others recognize and memorize which foreign agents have already breached our body so that the appropriate response can be sent the next time. Our body needs an optimal combination of each of these components. The clock genes play an important role in deciding how much of each type of immune cell our body should produce. When our clock system breaks down, it causes a cellular imbalance to our immune systems, producing more of one type of defense at the cost of another. For example, an imbalanced immune system that is good at destroying bacteria but not as great for wound repair will get exhausted from fighting new infections at a wound site. Or an immune system that cannot remember the last foreign agents it faced might produce a muted response to a new vaccine.

The circadian clock also regulates the basic defense mechanism inside every cell, regardless of whether the cell is part of any immune system. It's as if there is an immune system inside each cell that can neutralize threats. The most common threat inside a cell is *oxidative stress*, which occurs as a direct result of additional oxygen molecules entering the cell. These molecules produce dangerous free radicals, electronically unstable oxygen molecules that must scavenge electrons from whatever sources they can find in order to become stable molecules. The sources of electrons can include cellular DNA, cell membranes, important enzymes, and vital structural or functional proteins. When these important cell parts and substances lose their electrons and bind to free radicals, their function is altered.

Oxidative stress has been shown to be an important factor in many diseases because it leads to chronic and systemic inflammation. In fact, the number one biological mechanism that seems to underlie most chronic disease states is oxidative stress, and the results can include cancer, heart disease, dementia, arthritis, muscle damage, infection, and accelerated aging. One of the central roles of the circadian clock is to control oxidative stress. After eating, when every cell in our body uses nutrition to make energy, cells produce reactive oxygen species. The clock acts as a sensor of this oxidative state inside the cell and coordinates antioxidant defense mechanisms to clean up the damage. Since eating used to happen predictably during the daytime for millions of years, this function of the clock is very fundamental to cellular health. Scientists believe this predictable rise and fall of oxidative stress between day and night might have been one of the primary instigators of the evolution of the circadian clock.[2]

Another cellular activity is autophagy, the controlled digestion of bits and pieces of a cell, which helps to reduce some of the damage from oxidative stress. Let's say you are living in a remote town that does not provide garbage service to come pick up your trash. It's hard for you to get to the dump, so you try to recycle as much as possible, reusing items instead of throwing them away. Cells recycle their stray bits and pieces through autophagy as their internal immune system locates them and puts them in a garbage disposal system, which is called the *lysosome*. The lysosome contains acids that digest the cellular garbage. Once the cellular garbage is broken down, the raw material inside can be used again to build new cellular parts. Autophagy is more active several hours after our last meal (after several hours of fasting and before the first bite of the day), and then slows down when we eat. Time-restricted eating is known to increase autophagy for a few hours during the fasting period.[3]

Mitochondria are microscopic organelles that are found inside every cell, but more in muscle cells. They are the principal sites where all of our energy is generated. Faulty, damaged, or stressed mitochondria produce reactive oxygen species and autophagy cleans up both damaged mitochondria and other collateral damage inside the cell. A healthy circadian rhythm improves mitochondrial function, mitochondrial repair, and autophagy, which in turn improves overall cellular health.

Sometimes, autophagy and similar cleaning mechanisms may not be enough to neutralize cellular damage or stress. In this case, another layer of a more potent defense is triggered. This defense system can make any cell defend itself as if it is an immune cell. It lets the cell produce chemicals that can fight an infection or invite the tissue-resident immune cells to come to the rescue. Imagine this immune reaction within each cell as the fire alarm in your house. It is good to have it, but it is irritating and exhausting if your fire alarm is constantly on (chronic inflammation). Besides, when this cellular alarm system turns on—even during a real threat—it can take attention away from other functions of the cells. That is why chronic activation of this defense system can compromise the body's general functions, such as metabolism, damage repair, and more.[4]

We have found that when we disrupt the circadian code in mice, every cell behaves as if it is under attack.[5]

The Immune System's Circadian Response

Each immune system's distinct tasks—surveillance, attack, repair, and cleanup—occur on a schedule and at different times of the day. This may seem counterintuitive, because you might think that all immune responses should be happening at the same time: when the threat is detected. However, staggering the operations serves a very important lifesaving purpose. When multiple arms of our immune system are activated at once, it can overwhelm our body and cause a state of shock from which we cannot recover. This is called *septic shock*. By completing these chores at different times, the body can adjust more easily to the changes that are happening.

A large portion of the immune system is found in the gut, which is appropriate because the greatest number of potential invaders come in through the foods we eat or develop within the bacteria in our gut. As we discussed in Chapter 9, gut microbes flourish and deplete at different times of the day. When we lived in a less sanitized time, we were constantly exposed to bacteria, parasites, and viruses that would regularly set off our immune system. Many of these offending agents have their own circadian rhythms. In anticipation of the daily rise and fall of threats from bacteria and parasites, our immune system is programmed to have a circadian rhythm. This rhythm in immune function also serves as a check against chronic inflammation. In other words, the loss of circadian regulation of the immune system might be another cause of chronic inflammation.

Aside from the gut, there are other immune systems that are embedded in body fat, in our liver, and even in our brain. These immune systems act like security guards: Mostly they stand around waiting for something to happen. When an invader comes in, they become activated to neutralize the threat. For instance, if the gut is breached and bacterial particles get into the blood and reach these tissue-embedded immune cells, it can cause systemic inflammation. Circadian disruption also makes the tissue of brain cells stressed, and the stressed cells produce many chemicals that activate these tissue immune cells, leading to chronic inflammation.

Inflammation in the brain can contribute to depression, multiple sclerosis, and even schizophrenia. Inflammation of fat deposits is a general feature of obesity. This further compromises the fat cells' normal function to burn fat when needed. When the liver is damaged due to excessive fat deposits, it produces chemicals that invite immune cells to try to repair it. This leads to the liver filling up with scar tissue, also called steatohepatitis or in extreme cases liver cirrhosis.

Having a healthy circadian rhythm under TRE serves several important functions in reducing systemic inflammation. A strong circadian rhythm supports better repair of the skin and gut lining so that there is less opportunity for undigested food particles, disease-causing bacteria, or allergy-causing chemicals to enter our body and activate the immune system. A stronger circadian rhythm reduces oxidative stress and the production of inflammatory chemicals. With a reduction in external agents gaining access to our body and a reduction of our body's own inflammatory chemicals, the immune cells are less activated and thereby create less systemic inflammation under TRE.

Hacking Your Code Makes Recovery Easier

Even doctors will agree that the worst place for anyone to be is in a hospital, especially in the case of the elderly. It is quite common that patients with compromised immune systems pick up potentially deadly infections in the hospital. For instance, there is a recognized term, *Intensive Care Unit delirium,* or *ICU delirium,* which describes a state of impaired cognition that can contribute to long-term cognitive dysfunction.[6] Symptoms can include inattention, disturbance of consciousness within short periods, memory loss, confusion, and language or emotional disturbance.[7] ICU delirium can happen to anyone in a hospital room, given sleep deprivation and the lack of any sense of time or light. It's possible that ICU delirium occurs when the immune system is compromised, but we believe that it has more to do with a circadian disruption. When people are in the hospital, they are poked every 2 to 3 hours, they don't have continuous sleep, the lights are on all the time, and they are often attached to an IV line, which means food and medications are provided at random times or constantly.

The best defense in this case is a good offense: If you have to go to the hospital, make sure you have your best sleeping tools with you, especially a sleep mask and earplugs. One study investigated the effect of noise on the quality of sleep and the occurrence of ICU delirium. The use of earplugs at bedtime led to better sleep and delirium prevention, especially if used within 48 hours of admission.[8] In general, having a strong circadian rhythm helps you recover faster during a hospital stay because it improves tissue repair, reduces inflammation, helps regenerate damaged tissue, and minimizes stress on your body.

The Circadian Code for Taking Anti-Inflammatory Medications

If the body's inflammation process is circadian, you would expect that many inflammatory diseases may exacerbate at certain times of the day or night. For instance, one of the most widespread inflammatory diseases among older adults is arthritis, causing inflammation and severe pain in the joints. Many people with arthritis notice that the severity of pain and stiffness is at its peak in the morning, making it difficult to get out of bed.

Patients often take anti-inflammatory medications to control arthritis pain. In a study involving more than 500 patients suffering from rheumatoid arthritis, patients were given the popular nonsteroidal anti-inflammatory drug (NSAID) indomethacin in the morning, noon, or evening.[9] The incidence of related side effects, including stomach discomfort or headache/dizziness, was nearly fivefold after a morning dosing as compared with an evening dosing. The evening dosing also performed better in reducing the pain and stiffness that usually occur in the morning. It is known that inflammation that causes rheumatoid arthritis increases after midnight. So, taking any anti-inflammatory drug before bed can preemptively reduce the severity of nighttime inflammation, and you may wake up with less arthritis pain in the morning.

Steroid medications such as prednisone have strong anti-inflammatory effects. They work by slowing down your immune cells or suppressing their activity. We also know that our body's own steroids, such as cortisol, also rise slowly through nighttime, and arthritis patients produce less cortisol.[10] Therefore, scientists assumed that increasing steroid levels after midnight would be effective in combating arthritis. However, this regimen would be difficult to maintain, especially since we should all be fast asleep at midnight. The fix was the creation of an extended-release version of these drugs so that the patient can take it at bedtime, around 9:00 or 10:00 p.m., but the drug is released from capsules to the gut 3 to 4 hours later. A controlled clinical trial confirmed its efficacy: The study showed that when patients with rheumatoid arthritis were given the same dose of immediate-release prednisone at bedtime or an extended-release version, the late-night-release version had a 24 percent better reduction in morning stiffness of joints.[11]

In fact, scientists have found that the tolerability of nearly 500 medications improve by up to fivefold when they are matched to circadian scheduling.[12] The amount of every medication we take is balanced for two effects—the intended effect to treat the disease or symptom, and the unintended adverse side effects. That's why simply increasing the dose of a medication does not make treatment better or faster, as the adverse side effects may be too much to handle at a higher dose. Therefore, timing may be the answer to increasing efficacy. This may transform the treatment of many diseases, from cancer to high blood pressure, autoimmune diseases, heart diseases, depression, anxiety, and more.

When to Get a Flu Shot

Plan your vaccination day in advance, and try to get a week's worth of good sleep beforehand. In one study, when participants had poor sleep for a few days before vaccination, nearly half of them showed significantly delayed response to the vaccine.[13] This raises an important issue about the flu vaccine, as some people who are inoculated do not develop protection against the flu. These people might want to pay attention the next year and make sure that they slept well in the week prior to taking the flu shot.

In addition to sleep history, time of day appears to be another factor to consider while going for your flu shot. Preliminary studies have shown that morning vaccinations produce better protection than those given in the afternoon.[14]

TRE Helps Control Inflammation

Circadian rhythm disruption is known to compromise the immune system, leading to an increase in systemic inflammation and increased susceptibility to bacterial infection.[15] However, maintaining a robust circadian rhythm by TRE can help optimize immune function, reduce infection, and reduce systemic inflammation.[16] This may occur by multiple mechanisms.

We believe that this immune system benefit may be partly due to improved digestive health through TRE. As we learned in Chapter 9, when we improve the barrier function in the gut, fewer invaders will make their way into the bloodstream, and the circulatory immune cells will have fewer threats to neutralize. TRE also reduces systemic inflammation throughout the body, including in our fat stores. When our body fat is used as an energy source, the amount of inflammatory fat and general cellular damage is reduced. The reduction in inflammatory fat is being increasingly recognized as a contributing factor for staving off type 2 diabetes and insulin resistance. And as systemic inflammation reduces, joint pain and stiffness will go away, making physical activity possible and enjoyable. TRE also improves the brain clock, which reinforces a barrier—similar to the one found in the gut—that coats the brain so that only oxygenated blood can enter the brain, and not bacteria, cellular debris, or other particles that could compromise brain function. This can reduce local brain inflammation that causes many brain diseases, including dementia.

In addition, TRE also improves every cell's immune defense system. Our cells produce more antioxidants that neutralize free radicals when we follow TRE, so there is less cellular damage. TRE also improves autophagy, so more damage is removed and recycled. Finally, as the circadian clock inside the cell improves under TRE, it can tune the cell's own internal defense system for several hours every day. When our cells are healthy and less inflamed, the whole body works better.

Before I started to follow a TRE, I used to have sore knees and joints. Often, I had to wear knee braces or apply cold packs after exercise. When I traveled, I always got sick—catching a cold or developing an infection. All of my antibiotic prescriptions over the past several years have been for infections that developed following a few evenings of late-night work (and late-night snacks) or after transcontinental flights.

Since I started following TRE 6 years ago, I rarely get sick after travel, and I don't have joint pain; I have not used my knee braces or cold packs in years.

Cancer: The Ultimate Break in Circadian Rhythm

In 2007, the World Health Organization's International Agency for Research on Cancer declared shift work that involved circadian disruption to be a "probable" carcinogen. Over the past decade, additional research involving large longitudinal studies has extended the probable link between shift work and cancer to colorectal cancer, ovarian cancer, and breast cancer.

Cancer has many different causes, and some have a circadian component:

- Excessive inflammation: As we discussed, inflammation is circadian, and when chronic inflammation continues, particularly in the gut or liver, it contributes to cancer growth.

- Free radical oxidative stress: Free radicals can damage cellular DNA, and with the damaged DNA there are mutations, some of which can be cancerous.

- Telomeres: As the circadian clock is involved in DNA repair, it also has some effect on maintaining healthy telomeres (the ends of chromosomes). In one study, women who worked the night shift for 5 years or more had reduced telomere length and an associated increased risk for breast cancer.[17]

- Immune system surveillance: Some immune cells are on the lookout for tissue that doesn't look right, and when they find it, they kill it. This is a very clear example of productive autoimmunity because when the immune system finds a cancer cell that's 90 percent like a normal cell, it kills it. When this immune system is compromised, as happens under circadian disruption, many cancer cells escape a weak surveillance and grow to become life-threatening tumors.

- Cell cycle checkpoints: One of the fundamental differences between a normal cell and a cancer cell is that normal cells don't grow quickly nor do they divide that often, while cancer cells grow much faster and divide more often. When normal cells divide, they need to be in perfect form. The circadian clock in a normal cell makes sure that many control steps are in place for the cell to grow only at certain times, divide only once a day or every few days, and repair itself more regularly. Cancer cells escape all of these checks and balances. They grow much faster by escaping the circadian mechanism that rations nutrients for the cells. Cancer cells make more fat molecules that build new cells and they recycle their waste product to fuel their rapid growth. Cancer cells also don't have a stringent DNA damage repair mechanism, so they slowly accumulate DNA damage.

- Metabolism: Cells require a lot of energy when they are growing. The circadian clock controls metabolism, but when the clock is broken, metabolism speeds up, and that fuels cancer.

- DNA damage response: If DNA is damaged, it has to be repaired, and the circadian clock regulates some of the repair enzymes so that the repair system is on when the cells are likely to get damaged. For example, in the gut, the DNA repair system is on in the middle of the night. The skin's repair system is timed for the late evening so that it doesn't compete with daytime damage from the sun. If the timing for repair is off, the cell may divide into new cells before its damaged DNA is repaired. A proliferation of damaged DNA increases one's chance of developing cancer.

- Autophagy: Cancer cells use autophagy to fuel themselves. As soon as something is damaged, they immediately take it and recycle it. As we learned, autophagy is regulated by a clock, so it occurs only at certain times of the day, particularly in the middle of the night when we are fasting. When autophagy runs in high gear and does not have time to select all the damaged parts, sometimes damaged mitochondria are left behind. These damaged mitochondria, in turn, produce more free radicals or oxygen radicals or oxidative stress.

Cancer Treatment and Circadian Timing

Your circadian rhythm is relevant for many aspects of cancer, including prevention and treatment. The observation that shift workers with erratic timing of eating, sleeping, and exposure to light have increased risk for cancer immediately raises the possibility that maintaining a strong circadian rhythm will offer protection against cancer. In fact, in a large retrospective study on women and breast cancer risk, my colleague Ruth Patterson from the Moores Cancer Center at the University of California, San Diego, found that women who maintain a regular eating schedule and an 11-hour TRE are significantly protected from breast cancer.[18] Since TRE is known to reduce chronic inflammation—which is a recipe for cancer—it makes sense that TRE for 11 hours reduces breast cancer risk. This is a very important finding as there are very few studies linking nutrition to cancer risk that have been validated with independent controlled human studies.

Can simply changing daily habits reduce tumor growth as well? We think the answer may be yes, and the key is restoring circadian rhythm. A group of scientists tried this with mice and found positive results. They placed a tiny tumor in three groups of mice. The first group lived under a normal light-dark cycle, while the second group had their cycle changed every few days, as if they were experiencing jet lag or shift work. Both groups had access to food all the time and could eat whenever they wanted. They found that tumors grew more aggressively in mice under shift-work/jet-lag conditions. However, when mice in a third group were subjected to the same shift-work/jet-lag paradigm but were given access to food for only 12 hours, the tumor growth was reduced by as much as 20 percent in just 7 days.[19,20]

In terms of treatment, it has been known for more than 30 years that the timing of chemotherapy matters.[21] In one study that followed women with advanced ovarian cancer, patients were treated with two different drugs, doxorubicin and cisplatin, but at different times—a standard practice for ovarian cancer patients at that time. The women who took doxorubicin in the morning and cisplatin in the evening had less severe side effects from the cancer drugs, while the women who took the drugs on the opposite schedule—cisplatin in the morning and doxorubicin in the evening—had more severe side effects. This was the first study that showed that mistiming a drug led to worsened side effects. The study was reviewed in an article provocatively titled, "Dosing-Time Makes the Poison: Circadian Regulation and Pharmacotherapy."[22]

Since then, many studies with other types of cancers and different cancer drugs have shown the same conclusion—timing of cancer drugs can make the therapy less effective or more effective. In a study on colorectal cancer, the medication oxaliplatin was given to patients from a mini-pump that slowly delivered a small amount of the drug every hour, with a larger dose given at 4:00 p.m. Patients who had failed to respond to previous chemotherapy began responding positively to this timed dosing of cancer drug.[23]

Timing even comes into play when a tumor has to be removed. If the tumor has reached the liver, for example, nearly half of the liver containing the tumor is removed. After surgery, the normal liver cells are supposed to divide and grow so that the liver grows back to its normal size and carries out its normal function. In one study, a group of researchers in Japan removed two-thirds of the liver in mice in either the morning or in the late afternoon. Mice that underwent liver surgery in the afternoon showed much faster liver regrowth than mice that had surgery in the morning.[24]

Some cancer patients also have to endure total body irradiation (TBI) to destroy cancer cells in areas not easily reached by chemotherapy or surgery; this is typically deployed against cancer found in the nervous system, bones, skin, and, in men, the testes. Sometimes TBI is done to weaken or disable the immune system, especially if a patient is receiving a transplant. If a patient is getting bone marrow or stem cells from a donor, the patient's immune system sees these cells as foreign and tries to destroy them, thereby defeating the purpose of the treatment. TBI is also used to kill diseased bone marrow, so that new marrow will have space to grow. However, TBI has many adverse side effects, including hair loss, nausea, vomiting, and skin rash. This is because the radiation, intended to kill the cancer cells, also damages the DNA of normal cells; when the DNA is not repaired, the cells can die.

A few years ago, we did a simple experiment on laboratory mice. We found that the skin and hair cells of mice repair all damaged DNA in the evening. We took this finding one step further and tested what happens when TBI is given to mice at different times of the day. We treated one group of mice with an irradiation dose in the morning and another group with the same irradiation dose in the evening. The mice that received TBI in the morning, as expected, lost 80 percent of their hair. But the mice that received the same TBI in the evening retained 80 percent of their hair. This is because the irradiation given in the evening was in sync with their circadian clock, so the irradiation-induced DNA damage was promptly repaired and normal hair cell function was restored.[25]

The newest idea in cancer research and circadian rhythm is to develop drugs that will directly bind to clock molecules and restore circadian clock function in tumors, which naturally have poorly functioning clocks. Early research shows that brain cancer patients with normal levels of clock proteins in their tumor have longer survival rates than patients with low levels of clock proteins in their tumor.[26] In our lab, we reactivated a mouse tumor clock by treating glioblastoma with a drug that enhances the function of a clock gene.[27] When glioblastoma tumors were put into mice, the tumors grew aggressively, and within a few days the tumors increased in size by almost ten- or fifteenfold. But mice receiving the clock drug showed remarkably reduced tumor growth and also survived longer. More important, the clock drug was more effective than the standard drug used to treat brain cancer patients, which was given to a second group of mice.

Two Sisters Are Beating Cancer

Cancer treatment is complex. Even when cancer drugs kill some tumors, other tumors grow or after a few years of being cancer-free, dormant tumors may begin to appear. This is called *cancer recurrence*.

Synchronizing Patients and Caregivers

Once the causal connection between circadian rhythm and better cancer outcomes becomes widely acknowledged and accepted, doctors will shift their schedules to optimize for best outcomes. As you've seen, our circadian rhythm can be tweaked. For example, workers who are continuously on a night shift have a completely inverted circadian rhythm. Their melatonin level actually goes up during daytime and comes down during nighttime because they're living in a completely different time zone. Similarly, surgeons can invert their circadian rhythm so that they are at their peak performance when their patients' rhythms for treatment are also at their best. For instance, if surgery outcomes show better results in the afternoon, doctors can shift their peak performance from morning to afternoon as well by waking up late and working on an afternoon shift.

Advances in technology will continue to help us improve the treatment experience. For instance, in some European hospitals, the patient is hooked up to a mini-pump, like an insulin pump, that delivers medicines at the right time based on the individual's clock. Such technology can easily be used for many treatment options. Other treatments, including surgery, use robots via remote control: A doctor in New York can work on a patient in San Francisco or Hawaii via a robot.[28,29] This technological advance is another potential strategy for synchronizing the time lag between when the body is at its optimal time for surgery and when the doctor is in peak performance.

Cancer treatment research is strongly invested in understanding the circadian code, and in our lab, we are working on ways to connect the two. For instance, we are in touch with a pair of sisters: the older one has ovarian and uterine cancer, and the younger sister has breast cancer. Both sisters are following an 8-hour TRE and have reported that this eating schedule helps their treatment in many ways. They are experiencing less fatigue, less drug side effects such as nausea or intestinal pain, and better sleep. The TRE may even be boosting the efficacy of their cancer drugs. Their experience is in line with a recent study that also showed cancer recurrence was reduced among women who practice TRE.[30] TRE reduces the chance of tiny dormant tumor growth and thereby improves cancer treatment.

CHAPTER 12

The Circadian Code for Optimizing Brain Health

It's very difficult to know when the brain is not functioning well. We have an enormous capacity to compensate for our deficits, and we often think our behavior is normal, even when it isn't. Family and friends are often the first to notice a change in our behavior or thinking. And, when brain dysfunction begins in one family member, whether it occurs in the domain of thinking, emotional response, or memory, the whole family is affected. As dysfunction progresses, the sufferer may have difficulty maintaining normal family relations, be left alone with few or no friends, and can become a burden. Therefore, taking care of our mental health ensures that we are not only taking care of ourselves, but our family.

There is no single blood or genetic test that will tell us with absolute certainty that a person will have a lapse in brain function such as depression, anxiety, bipolar disorder, post-traumatic stress disorder (PTSD), or obsessive-compulsive disorder (OCD). What's more, at present, there is no cure for the diseases related to the brain such as Parkinson's, Alzheimer's, Huntington's, multiple sclerosis, and ALS (amyotrophic lateral sclerosis). These diseases may be linked to certain mutations in a few genes, yet this only explains a small fraction of all incidences, and it certainly doesn't justify the increase in incidences of many brain diseases in recent years.

It's more likely that genetic and environmental factors interact to produce disease. This is true for life-threatening brain disorders as well as depression, anxiety, and even OCD. And while this explanation makes sense, we do not know which specific environmental factors trigger these illnesses. However, we do know that maintaining a strong circadian rhythm builds resilience against these brain diseases.

What's the Circadian Part?

A circadian clock is present in almost all brain regions, including the areas that are implicated in neuropsychiatric diseases. Although we do not completely understand how brain dysfunction starts or develops, the mechanisms of these diseases primarily involve four themes, and the circadian clock is involved in all of them:

(1) Lack of emergence of new brain cells (neurons) that replace damaged or dead brain cells, causing a gradual decline in the number of healthy neurons: We used to believe that after our brain developed during childhood, we didn't make any more neurons. However, one of my colleagues at the Salk Institute, Fred Gage, exploded this notion almost 20 years ago.[1] It is now clear that adult brains have special stem cells that produce new neurons throughout our lifetime. These new neurons replace damaged or dead neurons through a process called *adult neurogenesis,* and the ability to regenerate is very important for maintaining a properly functioning brain well into old age. A reduced ability for neurogenesis contributes to a range of brain health dysfunction, from forgetfulness and memory loss to dementia.

The circadian clock regulates several aspects of adult neurogenesis. There is a daily pattern to the process of stem cells giving rise to new neurons, making sure that the right type of healthy fat molecules is delivered to the new neurons at the right time of the day. When we boost our circadian rhythm, more healthy neurons are created. Conversely, when we do not have enough sleep or when we experience jet lag, we reduce the number of new neurons that can be made that day.

(2) Poor wiring of neurons, causing misconnections/miscommunication between brain regions: Our brain is not fully developed when we are born. That means there are many parts of the brain that are not yet wired to other brain regions. These connections develop slowly over the first 5 years of life. Along with these connections comes a unique template of brain chemicals that mediate communications between neurons. During this critical period of development, sleep-wake and light-dark cycles affect brain development. An imbalance in the right amount of light (too little light during the day or too much at night) or an irregular sleep-wake cycle may leave a lasting impact in the form of a permanent change in sleep pattern, hypersensitivity to light, or even conditions like autism spectrum disorder (ASD) or attention deficit hyperactivity disorder (ADHD). In mice, melanopsin cells from the retina that are miswired to the brain can result in light-induced headache and migraine pain.[2] The same may be true in people when they spend too much time under bright light.

(3) The accumulation of damage, or lack of sufficient repair, and death of neurons: The circadian clock regulates the genes that are involved in reducing neuronal stress, promoting their repair so that neurons remain healthy. If any of the brain's circadian clocks are disrupted, the neurons associated with it can become easily stressed, damaged, or die off, or the process that cleans up the mess can be affected, causing more stress and damage. This is why a brain with a disrupted clock also has many missed connections. Misfiring chemicals in the brain cause further damage and can result in many different conditions, including autism, ADHD, depression, bipolar disorder, PTSD, generalized anxiety disorder, panic disorder, severe migraine, epilepsy, and seizures.

(4) A brain chemical imbalance: Neurons make brain chemicals called neurotransmitters that are the messengers between nerve cells. Some of these neurotransmitters include dopamine, serotonin, noradrenaline, and gamma-aminobutyric acid (GABA). These neurotransmitters mod-

ulate various aspects of brain function, including being alert or active, being calm, and responding to motivation or rewards. Many of these neurotransmitters are under circadian clock control, which makes sense because we typically go through different states of mind at different times of the day. In the morning, we are more alert and a little anxious about everything we have planned for the day. During the day, we are motivated to do what we have planned and are also driven by small rewards for getting stuff done. Toward the evening and late night, brain chemicals that support calmness help us to wind down.

Some brain clocks are involved in making these brain chemicals, while others are involved in breaking down the chemical production cycle. When the clock is disrupted, the daily rhythm of brain chemical production becomes mistimed or is stuck at high or low levels. This is when we develop different brain diseases. For example, when mice don't have a clock in their brain, they produce too much dopamine—a neurotransmitter related to energy use in the body, metabolism, and activity.[3] Too much dopamine makes both mice and men manic.

The Role of Light

The connection between our circadian code and mental health can be traced back to the migration of humans to the Northern Hemisphere 30,000 to 40,000 years ago. Lack of light is linked to depression, and short winter days with barely 6 hours of daylight are a prime culprit. Today, this is known as *seasonal affective disorder*—aptly abbreviated as SAD. SAD is a form of depression with symptoms including fatigue, hopelessness, and social withdrawal. Those who are susceptible to it feel the "winter blues" from the fall through the early spring, when people get less light naturally, and it does not resolve until the days become longer. This happens with frequency for those who live in Nordic countries during the winter, where the sun doesn't shine much in the morning hours while people are getting ready for work. As the day goes on and the sun comes out, mood elevates, and so does performance. Even outside of the northern European countries, depression and suicide rates within whole populations increase as we move from the equator to higher latitudes, and the increase is seasonal, with increased depression in winter.[4,5] This is a great example of an environmental factor that makes some communities susceptible to mental health issues.

A likely common thread between depression, seasonal affective disorder, and poor sleep at night (and feeling sleepy when we are awake) is the lack of sufficient bright light during the day. We know that those who have insomnia and feel sleepy throughout the day (irrespective of whether they have a regular job or shift work) are prone to depression.[6] However, we are just beginning to uncover the effect of light on lifelong sleep and activity patterns. A 2017 study showed that when adolescent mice are exposed to an unnatural day-night cycle that mimics a very gentle jet lag (traveling over one time zone), where the timing of light is advanced or delayed by as little as 1 hour each day for only a few weeks, their circadian clock completely rewires for the rest of the mouse's life.[7,8] We believe this effect is entirely due to how the master brain clock in the suprachiasmatic nucleus (SCN) reprograms itself by turning on or off a unique set of genes.

This research is groundbreaking because the type of circadian changes that appeared were previously believed only to occur when mice had a genetic mutation. The researchers found instead that these mice had a chemical imbalance in the SCN. The altered lighting schedule affected the production of the neurochemical GABA, which is known to keep us calm. Interestingly, the majority of SCN clock neurons produce GABA, and we also know that having too much or too little GABA has a huge impact on our daily organization of the sleep-wake cycle, as well as our ability to stay calm or become anxious.

Does this mean that children raised in a circadian-disruptive light environment are destined to mental health issues, or that adult habits can trigger brain dysfunction? We don't know for sure. But what we do know is that incidences of illness and disease related to brain dysfunction are on the rise. If we do a better job of setting a fixed bedtime, paying attention to how much light we are exposed to at night, and making sure we get more sunlight during the day, we might be able to reverse the course of these numbers.

Indoor lighting, especially at the wrong time, can have a profound effect on our circadian code, especially when we are sick. As discussed in chapter 11, it is known that many intensive care unit (ICU) patients, who are already in critical condition to begin with, experience a lack of a clear sense of day and night due to the fact that hospitals are always lighted. After a few days, many patients develop ICU delirium. Installing new lighting that simulates day-night change with bright and dim light, and reducing noise to support sleep at night, will restore circadian rhythm in ICU patients and considerably reduce ICU delirium.[9]

Premature babies are also exposed to the wrong types of light, and at the wrong time, from the day they are born. These babies come into this world when their brain and body are still developing and maturing. They spend their first days or weeks in a neonatal intensive care unit (NICU) until they are fully developed and can be raised at home. The lights in the NICU are always on because doctors and nurses have to check on the babies every few hours (if not minutes). There are also many monitors and computer screens that make noises and emit their own light. As a result, there is no sense of day or night for the developing child's brain. We also know that premature babies typically have many ongoing health problems, including brain development issues: Many grow up to have ADHD, ASD, learning disabilities, compromised language skills, and more. These correlative observations are raising new questions to test whether sustaining circadian rhythm by addressing exposure to light or the timing of eating can prevent or lessen the severity of these diseases.

In a very interesting study we discussed in Chapter 8, researchers safely covered premature infants' cribs for a few hours during the night, blocking bright light.[10] This simple adoption of a light-dark cycle accelerated the babies' growth and development to such an extent that it reduced their hospital stay by as much as 30 percent. The babies gained weight faster (a faster body weight gain correlates with better overall brain development) and their heart rate was more stable. Not only that, these babies had better oxygen saturation in their blood and had more melatonin. Simply letting them experience a clear day-night cycle had a profound effect.

The Right Light Beats Depression

Remember Cory Mapstone, the San Diego police officer we met in Chapter 7? When Cory works the night shift, he knows very well that he is prone to feeling low. But he has never had any spell of depression in 25 years of work. Why? He makes sure that he gets at least an hour of daylight exposure before he goes to bed. He told me that when he feels sunlight, it flows from his eyes to his brain and wakes it up. Being under daylight is like getting a free dose of a brain-lifting vitamin. Daylight rebalances brain chemicals— more light releases more excitatory glutamate in the brain, restores the daily rhythms of cortisol and melatonin, and maintains them in the right balance. Moreover, having more daylight also makes your sleep more resilient against the evening light so that you can still go to bed and beat anxiety.

Addressing Light and Sleep for Optimal Brain Health

A common theme of all neurological diseases is sleep disruption. Our daytime function is essentially a series of decisions that involve cognition and emotion. Sleep disruption affects this decision-making process. It is also commonly found in many psychiatric diseases, such as PTSD, anxiety, bipolar disorder, and more. It is also a prominent factor in neuro-degenerative diseases such as Alzheimer's disease and multiple sclerosis. These issues are rarely considered to be associated with abnormal sleep or circadian disruption, yet they should be.[11] Taking care of a sleep problem is typically a major aspect of treating any of these brain health problems.

Having too much exposure to light at night reduces sleep, which is when most of the cleanup of damaged cell proteins occurs. When you get more sleep, your brain naturally has more time to repair and cleanse its waste. Sleep also helps detox the brain in another way. In a newly discovered phenomenon, there seems to be a special drainage system in the brain, called the *brain lymphatic system*. This system operates during sleep to remove the metabolic waste of the brain. Sleep increases this process by as much as 60 percent.[12] So, whatever good habits you may have during the day, having a good night's sleep is the best way to remove all the waste products from your brain. This is presumed to prevent dementia.[13] An overstressed and sleep-deprived brain produces proteins that are not correctly shaped. As these misshaped proteins build up, they can cause death of the brain cells: a hallmark of dementia.

Does the Brain Forget How Much Sleep We Need as We Age?

As we've said, sleep is the time when memory is consolidated and stored. The more nights you can get up to 7 hours of sleep, the better protected you are from memory loss as you get older. This works in the short term, too: We learned in Chapter 4 that better sleep leads to better memory and attention the following day.

Some people ask me if their poor sleep habits will lead to long-term memory problems, such as dementia or Alzheimer's disease. The truth is, we don't know if poor sleep is a cause of dementia, but it is a contributor. Researchers have found that sleep deprivation impairs memory in mice and contributes to the development of the plaques and tangles that are the hallmark of Alzheimer's disease.[14,15] With this in mind, it's far better to have good sleep and protect your brain than to give up these hours unnecessarily.

Yet as we age, we are more likely to get less sleep, not more. Older adults have told me that they become wide awake after only 5 hours of sleep and feel refreshed enough so they don't attempt to go back to bed. We also know that the quality of sleep suffers as we get older. We are more sensitive to sound and light, both of which can disrupt sleep. Michael Rosbash (who received a 2017 Nobel Prize for discovering how our circadian clock works) and his team of researchers found that when they gently poked both young and old fruit flies, the older flies were more likely to wake up from their sleep.[16] The young flies went back to sleep or slept for a longer time the following day, as if they were making up for their lost sleep, while the older flies didn't sleep as long. It was as if their fruit fly brain "forgot" how much they were sleep-deprived. Rosbash's simple experiment suggests that an older brain not only wakes up with small disturbances, it even forgets how much time it needs to sleep.

Here's the takeaway: As we age, we do ourselves and our brains a disservice by shortchanging our opportunity for sleep. So, give yourself the chance to have 8 hours of sleep opportunity every night.

TRE for Brain Health

As we discussed in Chapter 9, hormones from the gut can mistakenly get into the blood. These hormones can cross into the brain and affect brain function. One of these hormones is CCK-4, which is known to cause anxiety if it reaches the brain. TRE will reduce the gut hormones that can act on your brain and cause anxiety or panic attack.

Another mechanism by which food affects brain function is when we focus too heavily on one nutrient group. For example, it has been known for almost a century that a ketogenic diet—one that is very low in carbs and high in fat—can reduce the incidence of severe drug-resistant childhood epilepsy. This nutrient profile changes the type of energy available to brain cells. Brain cells use ketone bodies (which are breakdown products of fat), which can improve the brain's overall function and reduce the incidence of seizures. An 8- to 10-hour TRE can also cause the body to tap into stored fat cells and produce these natural ketone bodies to use for brain energy. If you are doing an 8- to 10-hour TRE, for a few hours before breakfast, your body naturally produces ketones that will nurture your brain and reduce brain inflammation.

One of the most primitive responses is the motivation to look for food when food is scarce. We know that when mice are given food for only a few hours, they strategize and develop an interesting opportunistic food-seeking behavior.[17] They wake up a couple of hours before the food is supposed to come and start running around as if they are looking for food. There is evidence that these mice with limited food access use ketogenic energy in combination with their circadian clock to achieve the exact amount of sleep that would enable them to wake up early to look for food.[18]

There is also new evidence that ketones provide chemical signals that protect neurons from injury, or the neurons repair themselves better in the face of neurodegenerative diseases such as Alzheimer's, Parkinson's, and Huntington's diseases.[19] Although it is too early to connect the increase in motivation to look for food in animals going through TRE to improved brain health, it is now very clear that many benefits of a ketogenic diet on brain health can be tapped by eating all your food within 8 to 10 hours.

Eating at the same time every day and maintaining a long period without food synchronizes the circadian clocks in your brain and body. TRE naturally boosts the quality of your sleep, so you can sleep easily and uninterrupted for several hours.

In a 2018 study from the laboratory of Christopher Colwell at the University of California, Los Angeles, TRE was found to significantly reduce neurodegenerative symptoms of Huntington's disease in a mouse model.[20] Over 3 months, the mice with unlimited access to food developed the telltale signs of Huntington's disease: severe disruption of normal sleep-wake cycle, poor motor coordination, and increased heart-rate variability. The mice in the TRE group were significantly protected from these symptoms. The TRE mice slept well, their movement was better coordinated, their heart rate was more regular, and their brain function was more similar to that of a healthy brain.

Exercise Supports Brain Health

Exercise increases brain-derived neurotrophic factor (BDNF), which strengthens the connection between neurons and improves memory. BDNF can further augment repair of stressed or injured neurons—the process that also occurs when a strong circadian clock is present in the brain.

Both exercise and TRE can act independently or together to protect against the loss of dopaminergic neurons that happens in Parkinson's disease. The benefits are so profound that when mice that exercised were subjected to toxins that are known to kill neurons and cause Parkinson's disease, stroke, or even Huntington's disease, their brain was more resilient to these challenges and they recovered faster than mice that did not exercise.[21,22] Exercise or going 12 to 16 hours without food seems to produce similar chemical changes in the mouse brain[23] and can help maintain a robust circadian rhythm. These effects build resilience in our brain so that it can handle more damaging insults and can recover faster.

Dealing with Stress

Having a strong circadian clock is a protection against the stresses of everyday life that affect our health. For instance, the stress hormone cortisol is under strong circadian regulation. In healthy people, cortisol production peaks in the morning and reduces to a minimum level around bedtime. This allows us to wind down and go to sleep.

In a second mechanism, the circadian clock itself can negate the effect of a sudden spike in stress hormone production so that after the stressor is gone, we can return to a normal state of mind. Imagine you are stuck in traffic during your evening commute and you are running late to pick up your child from day care. Just worrying about being late is enough to increase the production of stress hormones from your adrenal gland. But when you finally arrive and are reunited with your child, even if you are late, your stress should dissipate; this calming down is related to a halt in the hormone production. The stress hormone that might have spiked during your commute and is circulating in your blood may not do much harm if your clock system is strong and was able to halt the stress hormone production.

But if you continue to feel stressed into the evening, you have a stress-response problem. First, too much stress hormone will overwhelm the system, and your own clock may not be able to handle it. The evening spike in stress hormone will keep you jazzed up. This delays your sleep time and may increase your exposure to nighttime bright light, which will further disrupt your clock. Some may think this "natural energy boost" is a positive experience because they feel they can be more productive at night. But, over time, constant energy at night can change from productivity into an anxiety disorder. And the next day, the fallout from this stress response can show up in many forms: If you went to sleep late, you may feel very tired, irritable, foggy, and hungry during the day.

People with chronic stress may even tip toward depression. Too much stress hormone throughout the day-night cycle reduces the creation of new neurons, and as the number of damaged neurons increases, we can succumb to depression. In older adults, the lack of new neurons in combination with increasing damaged or dead neurons can also contribute to forgetfulness and/or memory loss.

You can address your stress and fix it with good habits. Doing any exercise—even for 30 minutes to 1 hour in the gym—can buy you extra protection against the damaging effects of stress. In the evening, winding down with leisure reading or meditation helps reduce stress hormone production and promotes sleep.

Focusing on Depression

Stress or a sudden sad event can cause people to feel low, and it is natural for them to want to be alone, stay indoors, and brood in a dark room. All of these behaviors affect the circadian clock. And an affected clock pushes them further into depression. At the same time, one of the symptoms of depression is the inability to sleep or excessive sleep, which further disrupts the clock. Those who are already depressed may get into a vicious cycle. It's not the circadian clock causing depression, but depression causing a circadian disruption that in turn causes more depression.

One way to beat depression, or at least manage it, is to simplify your life in a very disciplined way. Good habits beget more good habits. If you can work on getting enough sleep at night, exercise during the daytime, increase your exposure to bright daylight, and have your meals at the same time every day, you can relieve some of life's stresses because these decisions have been made ahead of time.

Many stressful and unfortunate events are unavoidable. I have not met anyone who never had any stress or never faced a difficulty such as the loss of a job or a loved one. While these events can push us toward anxiety or depression, having a robust circadian clock is both a protection against and also a potential path out of these diseases.

You can maintain a robust clock and preserve your normal brain function by following four simple habits: sleep, TRE, exercise, and the appropriate exposure to daylight. Each of these four habits improves brain health. Improving one circadian habit gives you some benefit, but incorporating two or more goes a long way in nurturing your brain.

The vast majority of people with depression have difficulty falling asleep or maintaining sleep throughout the night. Many drugs used to treat depression act by promoting sleep. However, drug-induced sleep can make people too sleepy the next day, barely able to pull themselves out of bed. Although the drugs can slowly help them beat depression over weeks or months, quality of life is often compromised.

Others may experience a period of excessive alertness and activity, or mania. This is called *bipolar disorder*. It is now well documented that those with depression who also have irregular sleep patterns or get less sleep are more prone to mania, which can slowly progress to psychosis. Manic episodes may be triggered by traveling across several time zones coupled with less sleep.[24]

The connection between circadian disruption and brain diseases had long been thought to be correlative, yet it was difficult to prove how one actually caused the other. A few years ago, a direct connection was made between bipolar disorder and the circadian clock. One of the drugs used widely to treat bipolar disorders—lithium—was found to bind to one of the components of the circadian clock and make its function more potent.[25] This finding has implications that are both preventative and therapeutic in terms of brain health. In the same way, we know that people without depression sleep better and have better eating habits than those fighting depression. But you don't need lithium to achieve a positive mood; the circadian code habits of addressing sleep, light, food, and activity will help lift your spirits and improve brain health.

Roger Guillemin Credits His Longevity to His Rhythm

Roger Guillemin is a Nobel laureate as well as an artist, husband, father of six, and grandfather. But most impressively, he is still active and alert at 94 years old. In an interview with my postdoctoral researcher Emily Manoogian, he credited the routines he's established in his daily lifestyle (when, what, and how much he eats, sleeps, and moves) to part of his success.[26]

Dr. Guillemin grew up in Dijon, France, and stayed there to complete his bachelor's and begin his medical degree. He moved to Montreal to pursue his longtime interest in research and worked with Hans Selye, whose mentorship played a significant role in his life. Dr. Selye was one of the first scientists to understand the stress response, and how the adrenal glands produce chemicals like cortisol to help us cope in times of acute stress. As Dr. Guillemin explains, "Selye was the first to introduce the word *stress* into the medical conversation. Before that, *stress* was a term only used by engineers."

For 50 years as the head of his own lab, Dr. Guillemin's daily schedule was pretty consistent. (In fact, he has had the same assistant, Bernice, for more than 40 years.) He woke up around 6:30 to 7:00 a.m., without an alarm, every day. He was never much of a breakfast person, usually taking just a small meal: coffee with some toast and jam. He arrived at the lab around 8:00 a.m., sometimes had a small lunch around noon (no snacks), and then returned home after 5:00 p.m. to have dinner and a glass of wine with his family at 7:00 p.m. At home and whenever possible, he ate only French food. He never really restricted his diet but rather stuck to fresh quality food that he enjoyed. He went to bed around 10:00

p.m. and started again the next day. Dr. Guillemin never saw himself as an athlete, but he stayed active by swimming or playing tennis almost every day of his adult life.

In spite of his notable success, Dr. Guillemin experienced the common stresses on scientists to continually make progress in the lab. In fact, at times, he considered closing his lab. He did not have a special trick to relieve stress: He just remained persistent and noted that his daily routine and being with the family was, and continues to be, a wonderful support system throughout his life.

Maintaining optimal brain health is not necessarily a lifelong challenge, especially if you understand the central role of circadian rhythm in maintaining a healthy body and healthy mind. Just like the key to lifelong dental health is daily dental care, you can incorporate the simple habits we've outlined in this book to nurture your circadian rhythm. When you sleep, when you eat, and when you exercise can be orchestrated to the timely rhythms of your genes, hormones, and brain chemicals. In fact, every time I hear about someone like Dr. Guillemin living healthy into their nineties, a closer look at their lifestyle reveals they have incorporated circadian wisdom into their daily life.

CHAPTER 13

A Perfect Circadian Day

My perfect circadian days start the night before, when I finish dinner early—around 7:00 p.m.—and get to sleep by 10:30. In the morning, I feel rested and refreshed. After a hearty breakfast around 8:00 a.m., I take a quick walk outside, and then I get in my car and drive to work. My mind is buzzing during the drive, and by the time I get to my office I'm ready to get going. I take a short break for lunch around noon and then get back to work until 5:00 p.m. Then I get some exercise and head home to have dinner with my family and get some more work done or help my daughter with her homework after dinner, using task lighting.

As you've learned, these perfect days set my clock for optimal health. But do they happen every day? Of course not. I do a lot of traveling for my work; not just within the United States but all over the world. Sometimes I have to get up extra early to catch a plane or have a teleconference with colleagues a few time zones away. Sometimes I have to work late into the evening, staring at my computer into the night if I'm on a deadline. And sometimes I have to entertain colleagues or attend a conference that features a later dinner than I would like, throwing off my TRE.

However, I try my best every day to get as many aspects of my circadian code as right as possible. If I can't exercise, I make sure to hold fast to my TRE. If I eat a late dinner, I still try to give my stomach at least 12 to 13 hours rest before the next meal. If I go to sleep late, I make sure to exercise the next day. You get the idea. We're shooting for perfection,

Observed Benefits of Time-Restricted Eating

RANDOM EATING PATTERN	TIME-RESTRICTED EATING
Obesity	Reduced fat mass, increased lean mass
Glucose intolerance, insulin resistance	Normal glucose
High cholesterol	Normal cholesterol
Cardiovascular diseases	Improved cardiac function, reduced arrhythmia
Inflammation	Reduced inflammation
Fatty liver disease	Healthy liver
Increased cancer risk	Reduced cancer risk, better treatment outcomes
Poor sleep	Better sleep quality
Compromised muscle function	Increased endurance
Harmful gut microbiome	Healthy gut microbiome
Irregular bowel movement	Regular bowel movement
Kidney disease	Healthy kidney function
Poor motor coordination	Better motor coordination

but sometimes good enough has to do. I know that my health is in my hands: It's up to me to make the right choices as often as I can to reap the most rewards.

In reading this book, I hope that you've learned a thing or two about your circadian code and how easy it is to make the small changes necessary to strengthen it. After following the recommendations in this book for a few weeks, go back to the tests in Chapter 3 and see if your results are different. It's a great habit to keep track of the data you initially collected and see how you're doing in forming new habits. When you take your first bite and your last sets everything else in motion, especially if you can maintain an eating routine or schedule. Limiting light at night, especially exposure to bright lights, goes a long way to getting to sleep sooner and staying asleep longer. Exercise makes you tired and at the same time improves brain health—we know that the majority of work enhancing your brain health happens while you sleep anyway.

If you are currently suffering from chronic illness, remember that one of the best things you can do to reverse your course or lessen the severity is to enhance your circadian code. We are beginning to see many examples of people who are finding a whole new healthy lifestyle once they've tried these recommendations. Some have even reported to us that they no longer need to take medication. The role TRE plays in syncing your code and enhancing your health cannot be overstated. The chart on page 237 is meant to provide the best motivation to get you and your loved ones to adopt a TRE.

Enhancing your circadian code isn't a miracle cure, but at the same time, I hope you've learned that there is no magic in pills, either. By combining your doctor's recommendations with the information you've learned in this book, you will be doing everything in your power to get better and be healthy for life. My hope, of course, is that you will.

Acknowledgments

In June 2015 I was invited to an interdisciplinary science conference—Science Foo Camp—on the Google campus in Palo Alto. I gave my usual talk on circadian rhythms and their relevance to health, but this time a less academic audience seemed more intrigued than the PhDs at my usual lectures. These people, with diverse backgrounds and interests, wanted to know more about the hard science of circadian rhythms, and what people can implement now to improve their health and productivity. I realized that while there are many technical books written by eminent scientists in the field, there was no single book to convey this new science to a wider audience in order for people to use this information in their everyday lives.

One of the organizers and attendees of the conference, Linda Stone, had persistently encouraged me to write a book. The outline took shape from many dinner-table discussions with my family. My wife, Smita, and daughter, Sneha, would listen patiently to my scientific explanations and nudge me toward a simple clarification. Every once in a while, when my curious mother visited me, she would also join the discussion. My family's patience with my long hours in the lab and travel and their constant support have been priceless.

After seeing my presentation on circadian rhythms and health at the Near Future conference in March 2017, Maria Rodale invited me to write a book on circadian rhythms for the general public. The timing could not have been more perfect. I almost had what I thought was a good outline and content. But as soon as I started working on the book,

I realized I had to learn a completely new way to express my science. Pam Liflander came to my rescue and helped me arrange my thoughts and ideas clearly and coherently for anyone to get a simple and action- able message. My editors at Rodale, Marisa Vigilante, Shannon Welch, and Danielle Curtis, further refined the script and ensured that the readers would have access to the right references. Michael O'Conner went through the manuscript with a fine-tooth comb and provided an excellent copyedit. Lastly, Alyse Diamond at Penguin Random House brought the project to the finish line.

My scientific colleagues have been extremely helpful as well. In the first phase of my scientific career in circadian biology, my mentors were Steve Kay at the Scripps Research Institute and John Hogenesch at the Genomics Institute of the Novartis Research Foundation. Steve introduced me to the field of circadian biology and to many leaders in the field: I'm thrilled to know Jeff Hall, Michael Rosbash, and Michael Young, all of whom went on to win the Nobel Prize in Physiology and Medicine. They have inspired and influenced my research. I also drew inspiration from the basic science work of Susan Golden, Amita Sehgal, Jay Dunlap, and Takao Kondo. John Hogenesch played a catalyst role in my dive into circadian science relevant to human health. While at GNF, my collaborations with Joe Takahashi, Peter Schultz, Russ Van Gelder, Iggy Provencio, and Garret Fitzgerald led to many breakthrough discov- eries. These collaborations have continued, and both Steve and John have become my lifelong friends.

The next phase of my scientific career started when I joined the Salk Institute, where scientific excellence, symbiosis, and a strong drive to make foundational breakthroughs that can leave a lasting impact on the planet have been fueling my research. The work of the founder, Dr. Jonas Salk, is specifically inspiring for me: His invention of the polio vaccine proves the powerful message that prevention is the best cure. The Salk Institute has given me unwavering support to do many unconventional experiments. My principal collaborators and scientific colleagues at Salk include Ron Evans, Mark Montminy, Inder Verma, Rusty Gage, Martyn Goulding,

Reuben Shaw, and Joe Ecker, each of whom have helped me with my circadian-rhythm research in metabolism, neuroscience, epigenetics, rejuvenation, inflammation, and cancer. In addition, Kathy Jones and Joanne Chory have been constant sources of new ideas and direction.

Outside Salk, my collaborations and discussions with leaders in the area of metabolism and aging—Valter Longo, Mark Mattson, Leonard Guarente, and Johan Auwerx—helped integrate the science behind time-restricted eating and circadian rhythm with the science of longevity.

I am also truly fortunate to work with a great group of students and trainees. Their hard work and long hours in the lab breaking their own circadian code made it possible to test many of the ideas described in this book. I am especially thankful to Hiep Le, Nobushige Tanaka, Christopher Vollmers, Megumi Hatori, Shubhroz Gill, Amandine Chaix, Amir Zarrinpar, Ludovic Mure, Luciano DiTacchio, Masa Hirayama, Gabrielle Sulli, and Emily Manoogian. My countless discussions with Rosie Blau of *The Economist* and architect Frederick Marks were helpful in distilling how circadian lighting can be adopted in everyday life. I am also thankful to my physician friends Julie Wei-Shatzel, C. Michael Wright, and Pamela Taub, who have been guiding their patients with time-restricted eating.

I am also grateful to research funding from the National Institutes of Health, the Department of Defense, the Department of Homeland Security, the Leona M. and Harry B. Helmsley Charitable Trust, the Pew Charitable Trusts, the American Federation for Aging Research, the Glenn Foundation for Medical Research, the American Diabetes Association, World Cancer Research, the Joe W. and Dorothy Dorsett Brown Foundation, H. A. and Mary K. Chapman Charitable Trust, and Dr. and Mrs. Irwin and Joan Jacobs.

Finally, through the myCircadianClock.org website and research app, thousands of people have come to learn about their own circadian rhythms, and have shared their positive health changes achieved by following the lessons found in this book. I am grateful to all of them, especially the brave handful who have agreed to be included in this book.

NOTES

PREFACE

1 F. Damiola et al., "Restricted Feeding Uncouples Circadian Oscillators in Peripheral Tissues from the Central Pacemaker in the Suprachiasmatic Nucleus," *Genes and Development* 14 (2000): 2950–61.

2 K. A. Stokkan et al., "Entrainment of the Circadian Clock in the Liver by Feeding," *Science* 291 (2001) 490–93.

3 M. P. St-Onge, et al., "Meal Timing and Frequency: Implications for Cardiovascular Disease Prevention: A Scientific Statement from the American Heart Association," *Circulation* 135, no. 9 (2017): e96–e121.

CHAPTER I: WE ARE ALL SHIFT WORKERS

1 D. Fischer et al., "Chronotypes in the US—Influence of Age and Sex," *PLoS ONE* 12 (2017): e0178782.

2 T. Roenneberg et al., "Epidemiology of the Human Circadian Clock," *Sleep Medicine Reviews* 11, no. 6 (2007): 429–38.

3 L. Kaufman, "Your Schedule Could Be Killing You," *Popular Science*, September/October 2017, https://www.popsci.com/your-schedule-could-be-killing-you.

4 J. Li et al., "Parents' Nonstandard Work Schedules and Child Well-Being: A Critical Review of the Literature," *Journal of Primary Prevention* 35, no. 1 (2014): 53–73.

5 D. L. Brown et al., "Rotating Night Shift Work and the Risk of Ischemic Stroke," *American Journal of Epidemiology* 169, no. 11 (2009): 1370–77.

6 M. Conlon, N. Lightfoot, and N. Kreiger, "Rotating Shift Work and Risk of Prostate Cancer," *Epidemiology* 18, no. 1 (2007): 182–83.

7 S. Davis, D. K. Mirick, and R. G. Stevens, "Night Shift Work, Light at Night, and Risk of Breast Cancer," *Journal of the National Cancer Institute* 93, no. 20 (2001): 1557–62.

8 C. Hublin et al., "Shift-Work and Cardiovascular Disease: A Population-Based 22-Year Follow-Up Study," *European Journal of Epidemiology* 25, no. 5 (2010): 315–23.

9 B. Karlsson, A. Knutsson, and B. Lindahl, "Is There an Association between Shift Work and Having a Metabolic Syndrome? Results from a Population Based Study of 27,485 people," *Occupational & Environmental Medicine* 58, no. 11 (2001): 747–52.

10 T. A. Lahti et al., "Night-Time Work Predisposes to Non-Hodgkin Lymphoma," *International Journal of Cancer* 123, no. 9 (2008): 2148–51.

11 S. P. Megdal et al., "Night Work and Breast Cancer Risk: A Systematic Review and Meta-Analysis," *European Journal of Cancer* 41, no. 13 (2005): 2023–32.

12 F. A. Scheer et al., "Adverse Metabolic and Cardiovascular Consequences of Circadian Misalignment," *Proceedings of the National Academy of Sciences of the United States of America* 106, no. 11 (2009), 4453–58.

13 E. S. Schernhammer et al., "Night-Shift Work and Risk of Colorectal Cancer in the Nurses' Health Study," *Journal of the National Cancer Institute* 95, no. 11 (2003): 825–28.

14 E. S. Schernhammer et al., "Rotating Night Shifts and Risk of Breast Cancer in Women Participating in the Nurses' Health Study," *Journal of the National Cancer Institute* 93, no. 20 (2001): 1563–68.

15 S. Sookoian et al., "Effects of Rotating Shift Work on Biomarkers of Metabolic Syndrome and Inflammation," *Journal of Internal Medicine* 261, no. 3 (2007): 285–92.

16 A. N. Viswanathan, S. E. Hankinson, and E. S. Schernhammer, "Night Shift Work and the Risk of Endometrial Cancer," *Cancer Research* 67 no. 21 (2007): 10618–22.

17 E. S. Soteriades et al., "Obesity and Cardiovascular Disease Risk Factors in Firefighters: A Prospective Cohort Study," *Obesity Research* 13, no. 10 (2005): 1756–63.

18 E. S. Soteriades et al., "Cardiovascular Disease in US Firefighters: A Systematic Review," *Cardiology in Review* 19, no. 4 (2011): 202–15.

19 K. Straif et al., "Carcinogenicity of Shift-Work, Painting, and Fire-Fighting," *Lancet Oncology* 8, no. 12 (2007): 1065–66.

20 International Air Transport Association, "New Year's Day 2014 Marks 100 Years of Commercial Aviation," press release, http://www.iata.org/pressroom/pr/Pages/2013-12-30-01.aspx.

21 J.-J. de Mairan, "Observation Botanique," *Histoire de l'Academie Royale des Sciences* (1729): 35–36.

22 J. Aschoff, "Exogenous and endogenous components in circadian rhythms." *Cold Spring Harbor Symposia on Quantitative Biology* 25 (1960): 11–28.

23 J. Aschoff and R. Wever, "Spontanperiodik des Menschen bei Ausschluß aller Zeitgeber," *Naturwissenschaften* 49, no. 15 (1962): 337–42.

24 C. J. Morris, D. Aeschbach, and F. A. Scheer, "Circadian System, Sleep, and Endocrinology," *Molecular and Cellular Endocrinology* 349, no. 1 (2012): 91–104.

25 R. N. Carmody and R. W. Wrangham, "The Energetic Significance of Cooking," *Journal of Human Evolution* 57, no. 4 (2009): 379–91.

26 R. N. Carmody, G. S. Weintraub, and R. W. Wrangham, "Energetic Consequences of Thermal and Nonthermal Food Processing," *Proceedings of the National Academy of Sciences of the United States of America* 108, no. 48 (2011): 19199–203.

27 P. W. Wiessner, "Embers of Society: Firelight Talk among the Ju/'hoansi Bushmen," *Proceedings of the National Academy of Sciences of the United States of America* 111, no. 39 (2014): 14027–35.

28 R. Fouquet and P. J. G. Pearson, "Seven Centuries of Energy Services: The Price and Use of Light in the United Kingdom (1300–2000)," *Energy Journal* 27, no. 1 (2006): 139–77.

29 G. Yetish et al., "Natural Sleep and Its Seasonal Variations in Three Pre-Industrial Societies," *Current Biology* 25, no. 21 (2015): 2862–68.

30 H. O. de la Iglesia et al., "Ancestral Sleep," *Current Biology* 26, no. 7 (2016): R271–72.

31 H. O. de la Iglesia et al., "Access to Electric Light Is Associated with Shorter Sleep Duration in a Traditionally Hunter-Gatherer Community," *Journal of Biological Rhythms* 30, no. 4 (2015): 342–50.

32 R. G. Foster et al., "Circadian Photoreception in the Retinally Degenerate Mouse (rd/rd)," *Journal of Comparative Physiology A* 169, no. 1 (1991): 39–50.

33 M. S. Freeman et al., "Regulation of Mammalian Circadian Behavior by Non-Rod, Non-Cone, Ocular Photoreceptors," *Science* 284, no. 5413 (1999): 502–4.

34 R. J. Lucas et al., "Regulation of the Mammalian Pineal by Non-Rod, Non-Cone, Ocular Photoreceptors," *Science* 284, no. 5413 (1999): 505–7.

35 S. Panda et al., "Melanopsin (Opn4) Requirement for Normal Light-Induced Circadian Phase Shifting," *Science* 298, no. 5601 (2002): 2213–16.

36 N. F. Ruby et al., "Role of Melanopsin in Circadian Responses to Light," *Science* 298, no. 5601 (2002): 2211–13.

37 S. Hattar et al., "Melanopsin-Containing Retinal Ganglion Cells: Architecture, Projections, and Intrinsic Photosensitivity," *Science* 295, no. 5557 (2002): 1065–70.

38 D. M. Berson, F. A. Dunn, and M. Takao, "Phototransduction by Retinal Ganglion Cells That Set the Circadian Clock," *Science* 295, no. 5557 (2002): 1070–73.

39 I. Provencio et al., "Melanopsin: An Opsin in Melanophores, Brain, and Eye," *Proceedings of the National Academy of Sciences of the United States of America* 95, no. 1 (1998): 340–45.

CHAPTER 2: HOW CIRCADIAN RHYTHMS WORK: TIMING IS EVERYTHING

1 R. J. Konopka and S. Benzer, "Clock Mutants of *Drosophila melanogaster*," *Proceedings of the National Academy of Sciences of the United States of America* 68, no. 9 (1971): 2112–16.

2 S. Panda et al., "Coordinated Transcription of Key Pathways in the Mouse by the Circadian Clock," *Cell* 109, no. 3 (2002): 307–20.

3 D. K. Welsh, J. S. Takahashi, and S. A. Kay, "Suprachiasmatic Nucleus: Cell Autonomy and Network Properties," *Annual Review of Physiology* 72 (2010): 551–77.

4 R. E. Fargason et al., "Correcting Delayed Circadian Phase with Bright Light Therapy Predicts Improvement in ADHD Symptoms: A Pilot Study," *Journal of Psychiatric Research* 91 (2017): 105–10.

5 T. Roenneberg et al., "Epidemiology of the Human Circadian Clock," *Sleep Medicine Reviews* 11, no. 6 (2007): 429–38.

6 K. L. Toh et al., "An hPer2 Phosphorylation Site Mutation in Familial Advanced Sleep Phase Syndrome," *Science* 291, no. 5506 (2001): 1040–43.

7 Y. He et al., "The Transcriptional Repressor DEC2 Regulates Sleep Length in Mammals," *Science* 325, no. 5942 (2009): 866–70.

8 K. P. Wright, Jr. et al., "Entrainment of the Human Circadian Clock to the Natural Light-Dark Cycle," *Current Biology* 23, no. 16 (2013): 1554–58.

9 C. Vollmers et al., "Time of Feeding and the Intrinsic Circadian Clock Drive Rhythms in Hepatic Gene Expression," *Proceedings of the National Academy of Sciences of the United States of America* 106, no. 50 (2009): 21453–58.

10 D. M. Edgar et al., "Influence of Running Wheel Activity on Free-Running Sleep/ Wake and Drinking Circadian Rhythms in Mice," *Physiology & Behavior* 50, no. 2 (1991): 373–78.

11 S. Brand et al., "High Exercise Levels Are Related to Favorable Sleep Patterns and Psychological Functioning in Adolescents: A Comparison of Athletes and Controls," *Journal of Adolescent Health* 46, no. 2 (2010): 133–41.

12 K. J. Reid et al., "Aerobic Exercise Improves Self-Reported Sleep and Quality of Life in Older Adults with Insomnia," *Sleep Medicine* 11, no. 9 (2010): 934–40.

13 S. S. Tworoger et al., "Effects of a Yearlong Moderate-Intensity Exercise and a Stretching Intervention on Sleep Quality in Postmenopausal Women," *Sleep* 26, no. 7 (2003): 830–36.

14 E. J. van Someren et al., "Long-Term Fitness Training Improves the Circadian Rest-Activity Rhythm in Healthy Elderly Males," *Journal of Biological Rhythms* 12, no. 2 (1997): 146–56.

CHAPTER 3: TRACK AND TEST: IS YOUR CIRCADIAN CODE IN SYNC?

1 F. C. Bell and M. L. Miller, "Life Tables for the United States Social Security Area 1900–2100," Social Security Administration, https://www.ssa.gov/oact/NOTES/as120/LifeTables_Body.html.

2 C. R. Marinac et al., "Prolonged Nightly Fasting and Breast Cancer Prognosis," *JAMA Oncology* 2, no. 8 (2016): 1049–55.

3 A. J. Davidson et al., "Chronic Jet-Lag Increases Mortality in Aged Mice," *Current Biology* 16, no. 21 (2006): R914–16.

4 D. C. Mohren et al., "Prevalence of Common Infections Among Employees in Different Work Schedules," *Journal of Occupational and Environmental Medicine* 44, no. 11 (2002): 1003–11.

5 N. J. Schork, "Personalized Medicine: Time for One-Person Trials," *Nature* 520, no. 7549 (2015): 609–11.

6 B. J. Hahm et al., "Bedtime Misalignment and Progression of Breast Cancer," *Chronobiology International* 31, no. 2 (2014): 214–21.

7 E. L. McGlinchey et al., "The Effect of Sleep Deprivation on Vocal Expression of Emotion in Adolescents and Adults," *Sleep* 34, no. 9 (2011): 1233–41.

8 S. J. Wilson et al., "Shortened Sleep Fuels Inflammatory Responses to Marital Conflict: Emotion Regulation Matters," *Psychoneuroendocrinology* 79 (2017): 74–83.

9 S. Gill and S. Panda, "A Smartphone App Reveals Erratic Diurnal Eating Patterns in Humans That Can Be Modulated for Health Benefits," *Cell Metabolism* 22, no. 5 (2015): 789–98.

10 Ibid.

11 N. J. Gupta, V. Kumar, and S. Panda, "A Camera-Phone Based Study Reveals Erratic Eating Pattern and Disrupted Daily Eating-Fasting Cycle among Adults in India," *PLoS ONE* 12, no. 3 (2017): e0172852.

12 M. Ohayon et al., "National Sleep Foundation's Sleep Quality Recommendations: First Report," *Sleep Health* 3, no. 1 (2017): 6–19.

13 M. Hirshkowitz et al., "National Sleep Foundation's Sleep Time Duration Recommendations: Methodology and Results Summary," *Sleep Health* 1, no. 1 (2015): 40–43.

14 M. Hirshkowitz et al., "National Sleep Foundation's Updated Sleep Duration Recommendations: Final Report," *Sleep Health* 1, no. 4 (2015): 233–43.

CHAPTER 4: A CIRCADIAN CODE FOR THE BEST NIGHT'S SLEEP

1 M. Hirshkowitz et al., "National Sleep Foundation's Sleep Time Duration Recommendations: Methodology and Results Summary," *Sleep Health* 1, no. 1 (2015): 40–43.

2 M. Hirshkowitz et al., "National Sleep Foundation's Updated Sleep Duration Recommendations: Final Report," *Sleep Health* 1, no. 4 (2015): 233–43.

3 D. F. Kripke et al., "Mortality Associated with Sleep Duration and Insomnia," *Archives of General Psychiatry* 59, no. 2 (2002): 131–36.

4 G. Yetish et al., "Natural Sleep and Its Seasonal Variations in Three Pre-Industrial Societies," *Current Biology* 25, no. 21 (2015): 2862–68.

5 H. O. de la Iglesia et al., "Access to Electric Light Is Associated with Shorter Sleep Duration in a Traditionally Hunter-Gatherer Community," *Journal of Biological Rhythms* 30, no. 4 (2015): 342–50.

6 A. M. Williamson and A. M. Feyer, "Moderate Sleep Deprivation Produces Impairments in Cognitive and Motor Performance Equivalent to Legally Prescribed Levels of Alcohol Intoxication," *Occupational & Environmental Medicine* 57, no. 10 (2000): 649–55.

7 H. P. van Dongen et al., "The Cumulative Cost of Additional Wakefulness: Dose-Response Effects on Neurobehavioral Functions and Sleep Physiology from Chronic Sleep Restriction and Total Sleep Deprivation," *Sleep* 26, no. 2 (2003): 117–26.

8 R. E. Fargason et al., "Correcting Delayed Circadian Phase with Bright Light Therapy Predicts Improvement in ADHD Symptoms: A Pilot Study," *Journal of Psychiatric Research* 91 (2017): 105–10.

9 N. Kronfeld-Schor and H. Einat, "Circadian Rhythms and Depression: Human Psychopathology and Animal Models," *Neuropharmacology* 62, no. 1 (2012): 101–14.

10 M. E. Coles, J. R. Schubert, and J. A. Nota, "Sleep, Circadian Rhythms, and Anxious Traits," *Current Psychiatry Reports* 17, no. 9 (2015): 73.

11 S. E. Anderson et al., "Self-Regulation and Household Routines at Age Three and Obesity at Age Eleven: Longitudinal Analysis of the UK Millennium Cohort Study," *International Journal of Obesity* 41, no. 10 (2017): 1459–66.

12 A. W. McHill et al., "Impact of Circadian Misalignment on Energy Metabolism during Simulated Nightshift Work," *Proceedings of the National Academy of Sciences of the United States of America* 111, no. 48 (2014): 17302–7.

13 B. Martin, M. P. Mattson, and S. Maudsley, "Caloric Restriction and Intermittent Fasting: Two Potential Diets for Successful Brain Aging," *Ageing Research Reviews* 5, no. 3 (2006): 332–53.

14 S. Gill and S. Panda, "A Smartphone App Reveals Erratic Diurnal Eating Patterns in Humans That Can Be Modulated for Health Benefits," *Cell Metabolism* 22, no. 5 (2015): 789–98.

15 S. J. Crowley and C. I. Eastman, "Human Adolescent Phase Response Curves to Bright White Light," *Journal of Biological Rhythms* 32, no. 4 (2017): 334–44.

16 J. A. Evans et al., "Dim Nighttime Illumination Alters Photoperiodic Responses of Hamsters through the Intergeniculate Leaflet and Other Photic Pathways," *Neuroscience* 202 (2012): 300–308.

17 L. S. Gaspar et al., "Obstructive Sleep Apnea and Hallmarks of Aging," *Trends in Molecular Medicine* 23, no. 8 (2017): 675–92.

18 E. Ferracioli-Oda, A. Qawasmi, and M. H. Bloch, "Meta-Analysis: Melatonin for the Treatment of Primary Sleep Disorders," *PLoS ONE* 8, no. 5 (2013): e63773.

CHAPTER 5: TIME-RESTRICTED EATING: SET YOUR CLOCK FOR WEIGHT LOSS

1 C. M. McCay and M. F. Crowell, "Prolonging the Life Span," *Scientific Monthly* 39, no. 5 (1934): 405–14.

2 S. K. Das, P. Balasubramanian, and Y. K. Weerasekara, "Nutrition Modulation of Human Aging: The Calorie Restriction Paradigm," *Molecular and Cellular Endocrinology* 455 (2017): 148–57.

3 A. Kohsaka et al., "High-Fat Diet Disrupts Behavioral and Molecular Circadian Rhythms in Mice," *Cell Metabolism* 6, no. 5 (2007): 414–21.

4 M. Hatori et al., "Time-Restricted Feeding without Reducing Caloric Intake Prevents Metabolic Diseases in Mice Fed a High-Fat Diet," *Cell Metabolism* 15, no. 6 (2012): 848–60.

5 A. Chaix et al., "Time-Restricted Feeding Is a Preventative and Therapeutic Intervention against Diverse Nutritional Challenges," *Cell Metabolism* 20, no. 6 (2014): 991–1005.

6 A. Zarrinpar et al., "Diet and Feeding Pattern Affect the Diurnal Dynamics of the Gut Microbiome," *Cell Metabolism* 20, no. 6 (2014): 1006–17.

7 V. A. Acosta-Rodriguez et al., "Mice under Caloric Restriction Self-Impose a Temporal Restriction of Food Intake as Revealed by an Automated Feeder System," *Cell Metabolism* 26, no. 1 (2017): 267–77.e2.

8 M. Garaulet et al., "Timing of Food Intake Predicts Weight Loss Effectiveness," *International Journal of Obesity* 37, no. 4 (2013): 604–11.

9 S. Gill and S. Panda, "A Smartphone App Reveals Erratic Diurnal Eating Patterns in Humans That Can Be Modulated for Health Benefits," *Cell Metabolism* 22, no. 5 (2015): 789–98.

10 T. Moro et al., "Effects of Eight Weeks of Time-Restricted Feeding (16/8) on Basal Metabolism, Maximal Strength, Body Composition, Inflammation, and Cardiovas-

cular Risk Factors in Resistance-Trained Males," *Journal of Translational Medicine* 14 (2016): 290.

11 J. Rothschild et al., "Time-Restricted Feeding and Risk of Metabolic Disease: A Review of Human and Animal Studies," *Nutrition Reviews* 72, no. 5 (2014): 308–18.

12 T. Ruiz-Lozano et al., "Timing of Food Intake Is Associated with Weight Loss Evolution in Severe Obese Patients after Bariatric Surgery," *Clinical Nutrition* 35, no. 6 (2016): 1308–14.

13 A. W. McHill et al., "Later Circadian Timing of Food Intake Is Associated with Increased Body Fat," *American Journal of Clinical Nutrition* 106, no. 6 (2017): 1213–19.

14 National Institute of Diabetes and Digestive and Kidney Diseases, "Digestive Diseases Statistics for the United States," https://www.niddk.nih.gov/health-information/health-statistics/digestive-diseases.

15 McHill, "Later Circadian Timing."

16 J. Suez et al., "Artificial Sweeteners Induce Glucose Intolerance by Altering the Gut Microbiota," *Nature* 514, no. 7521 (2014): 181–86.

CHAPTER 6: OPTIMIZING LEARNING AND WORKING

1 J. S. Durmer and D. F. Dinges, "Neurocognitive Consequences of Sleep Deprivation," *Seminars in Neurology* 25, no. 1 (2005): 117–29.

2 S. M. Greer, A. N. Goldstein, and M. P. Walker, "The Impact of Sleep Deprivation on Food Desire in the Human Brain," *Nature Communications* 4 (2013). article no. 2259.

3 R. Stickgold, "Sleep-Dependent Memory Consolidation," *Nature* 437, no. 7063 (2005): 1272–78.

4 T. A. LeGates et al., "Aberrant Light Directly Impairs Mood and Learning through Melanopsin-Expressing Neurons," *Nature* 491, no. 7425 (2012): 594–98.

5 M. Boubekri, et al., "Impact of Windows and Daylight Exposure on Overall Health and Sleep Quality of Office Workers: A Case-Control Pilot Study," *Journal of Clinical Sleep Medicine* 10, no. 6 (2014): 603–11.

6 P. Meerlo, A. Sgoifo, and D. Suchecki, "Restricted and Disrupted Sleep: Effects on Autonomic Function, Neuroendocrine Stress Systems and Stress Responsivity," *Sleep Medicine Reviews* 12, no. 3 (2008): 197–210.

7 J. A. Foster and K. A. McVey Neufeld, "Gut-Brain Axis: How the Microbiome Influences Anxiety and Depression," *Trends in Neurosciences* 36, no. 5 (2013): 305–12.

8 S. J. Kentish and A. J. Page, "Plasticity of Gastro-Intestinal Vagal Afferent Endings," *Physiology & Behavior* 136 (2014): 170–78.

9 L. A. Reyner et al., "'Post-Lunch' Sleepiness During Prolonged, Monotonous Driving—Effects of Meal Size," *Physiology & Behavior* 105, no. 4 (2012): 1088–91.

10 M. S. Ganio, et al., "Mild Dehydration Impairs Cognitive Performance and Mood of Men," *British Journal of Nutrition* 106, no. 10 (2011): 1535–43.

11 T. Partonen and J. Lönnqvist, "Bright Light Improves Vitality and Alleviates Distress in Healthy People," *Journal of Affective Disorders* 57, no. 1–3 (2000): 55–61.

12 D. H. Avery et al., "Bright Light Therapy of Subsyndromal Seasonal Affective Disorder in the Workplace: Morning vs. Afternoon Exposure," *Acta Psychiatrica Scandinavica* 103, no. 4 (2001): 267–74.

13 C. Cajochen et al., "Evening Exposure to a Light-Emitting Diodes (LED)-Backlit Computer Screen Affects Circadian Physiology and Cognitive Performance," *Journal of Applied Physioliology* 110, no. 5 (2011): 1432–38.

14 A. M. Chang et al., "Evening Use of Light-Emitting eReaders Negatively Affects Sleep, Circadian Timing, and Next-Morning Alertness," *Proceedings of the National Academy of Sciences of the United States of America* 112, no. 4 (2015): 1232–37.

15 M. P. Mattson and R. Wan, "Beneficial Effects of Intermittent Fasting and Caloric Restriction on the Cardiovascular and Cerebrovascular Systems," *Journal of Nutritional Biochemistry* 16, no. 3 (2005): 129–37.

16 R. K. Dishman et al., "Neurobiology of Exercise," *Obesity* 14, no. 3 (2006): 345–56.

17 E. Guallar, "Coffee Gets a Clean Bill of Health," *BMJ* 359 (2017): j5356.

18 R. Poole et al., "Coffee Consumption and Health: Umbrella Review of Meta-Analyses of Multiple Health Outcomes," *BMJ* 359 (2017): j5024.

19 I. Clark and H. P. Landolt, "Coffee, Caffeine, and Sleep: A Systematic Review of Epidemiological Studies and Randomized Controlled Trials," *Sleep Medicine Reviews* 31 (2017): 70–78.

20 J. Shearer and T. E. Graham, "Performance Effects and Metabolic Consequences of Caffeine and Caffeinated Energy Drink Consumption on Glucose Disposal," *Nutrition Reviews* 72, Suppl. 1 (2014): 121–36.

21 T. M. Burke et al., "Effects of Caffeine on the Human Circadian Clock In Vivo and In Vitro," *Science Translational Medicine* 7, no. 35 (2015): 305ra146.

22 S. Grossman, "These Are the Most Popular Starbucks Drinks Across the U.S.," *Time*, July 1, 2014.

23 H. P. van Dongen and D. F. Dinges, "Sleep, Circadian Rhythms, and Psychomotor Vigilance," *Clinics in Sports Medicine* 24, no. 2 (2005): 237–49.

24 B. L. Smarr, "Digital Sleep Logs Reveal Potential Impacts of Modern Temporal Structure on Class Performance in Different Chronotypes," *Journal of Biological Rhythms* 30, no. 1 (2015): 61–67.

25 K. Wahlstrom, "Changing Times: Findings from the First Longitudinal Study of Later High School Start Times," *National Association of Secondary School Principals Bulletin* 86, no. 633 (2002): 3–21.

26 J. Boergers, C. J. Gable, and J. A. Owens, "Later School Start Time Is Associated with Improved Sleep and Daytime Functioning in Adolescents," *Journal of Developmental and Behavioral Pediatrics* 35, no. 1 (2014): 11–17.

27 J. A. Owens, K. Belon, and P. Moss, "Impact of Delaying School Start Time on Adolescent Sleep, Mood, and Behavior," *Archives of Pediatric & Adolescent Medicine* 164, no. 7 (2010): 608–14.

CHAPTER 7: SYNCING YOUR EXERCISE TO
YOUR CIRCADIAN CODE

1 M. S. Tremblay et al., "Physiological and Health Implications of a Sedentary Life-style," *Applied Physiology, Nutrition, and Metabolism* 35, no. 6 (2010): 725–40.

2 T. Althoff et al., "Large-Scale Physical Activity Data Reveal Worldwide Activity Inequality," *Nature* 547, no. 7663 (2017): 336–39.

3 D. R. Bassett, P. L. Schneider, and G. E. Huntington, "Physical Activity in an Old Order Amish Community," *Medicine and Science in Sports and Exercise* 36, no. 1 (2004): 79–85.

4 H. O. de la Iglesia et al., "Access to Electric Light Is Associated with Shorter Sleep Duration in a Traditionally Hunter-Gatherer Community," *Journal of Biological Rhythms* 30, no. 4 (2015): 342–50.

5 T. Kubota et al., "Interleukin-15 and Interleukin-2 Enhance Non-REM Sleep in Rabbits," *American Journal of Physiology: Regulatory Integrative and Comparative Physiology* 281, no. 3 (2001): R1004–12.

6 Y. Li et al., "Association of Serum Irisin Concentrations with the Presence and Severity of Obstructive Sleep Apnea Syndrome," *Journal of Clinical Laboratory Analysis* 31, no. 5 (2016): e22077.

7 K. M. Awad et al., "Exercise Is Associated with a Reduced Incidence of Sleep-Disordered Breathing," *American Journal of Medicine* 125, no. 5 (2012): 485–90.

8 J. C. Ehlen et al., "*Bmal1* Function in Skeletal Muscle Regulates Sleep," *eLife* 6 (2017): e26557.

9 E. Steidle-Kloc et al., "Does Exercise Training Impact Clock Genes in Patients with Coronary Artery Disease and Type 2 Diabetes Mellitus?" *European Journal of Preventive Cardiology* 23, no. 13 (2016): 1375–82.

10 N. Yang, and Q. J. Meng, "Circadian Clocks in Articular Cartilage and Bone: A Compass in the Sea of Matrices," *Journal of Biological Rhythms* 31, no. 5 (2016): 415–27.

11 E. A. Schroder et al., "Intrinsic Muscle Clock Is Necessary for Musculoskeletal Health," *Journal of Physiology* 593, no. 24 (2015): 5387–404.

12 S. Aoyama and S. Shibata, "The Role of Circadian Rhythms in Muscular and Osseous Physiology and Their Regulation by Nutrition and Exercise," *Frontiers in Neuroscience* 11 (2017): article no. 63.

13 E. Woldt et al., "Rev-erb-α Modulates Skeletal Muscle Oxidative Capacity by Regulating Mitochondrial Biogenesis and Autophagy," *Nature Medicine* 19, no. 8 (2013): 1039–46.

14 H. van Praag et al., "Running Enhances Neurogenesis, Learning, and Long-Term Potentiation in Mice," *Proceedings of the National Academy of Sciences of the United States of America* 96, no. 23 (1999): 13427–31.

15 J. L. Yang et al., "BDNF and Exercise Enhance Neuronal DNA Repair by Stimulating CREB-Mediated Production of Apurinic/Apyrimidinic Endonuclease 1," *NeuroMolecular Medicine* 16, no. 1 (2014): 161–74.

16 S. M. Nigam et al., "Exercise and BDNF Reduce Aβ Production by Enhancing A-Secretase Processing of APP," *Journal of Neurochemistry* 142, no. 2 (2017): 286–96.

17 W. D. van Marken Lichtenbelt et al., "Cold-Activated Brown Adipose Tissue in Healthy Men," *New England Journal of Medicine* 360, no. 15 (2009): 1500–1508.

18 V. Ouellet et al., "Brown Adipose Tissue Oxidative Metabolism Contributes to Energy Expenditure During Acute Cold Exposure in Humans," *Journal of Clinical Investigation* 122, no. 2 (2012): 545–52.

19 E. Thun et al., "Sleep, Circadian Rhythms, and Athletic Performance," *Sleep Medicine Reviews* 23 (2015): 1–9.

20 E. Facer-Childs and R. Brandstaetter, "The Impact of Circadian Phenotype and Time Since Awakening on Diurnal Performance in Athletes," *Current Biology* 25, no. 4 (2015): 518–22.

21 R. S. Smith, C. Guilleminault, and B. Efron, "Circadian Rhythms and Enhanced Athletic Performance in the National Football League," *Sleep* 20, no. 5 (1997): 362–65.

22 N. A. King, V. J. Burley, and J. E. Blundell, "Exercise-Induced Suppression of Appetite: Effects on Food Intake and Implications for Energy Balance," *European Journal of Clinical Nutrition* 48, no. 10 (1994): 715–24.

23 E. A. Richter and M. Hargreaves, "Exercise, GLUT4, and Skeletal Muscle Glucose Uptake," *Physiological Reviews* 93, no. 3 (2013): 993–1017.

24 E. van Cauter et al., "Nocturnal Decrease in Glucose Tolerance during Constant Glucose Infusion," *Journal of Clinical Endocrinology and Metabolism* 69, no. 3 (189): 604–11.

25 J. Sturis et al., "24-Hour Glucose Profiles during Continuous or Oscillatory Insulin Infusion: Demonstration of the Functional Significance of Ultradian Insulin Oscillations," *Journal of Clinical Investigation* 95, no. 4 (1995): 1464–71.

26 H. H. Fullagar et al., "Sleep and Athletic Performance: The Effects of Sleep Loss on Exercise Performance, and Physiological and Cognitive Responses to Exercise," *Sports Medicine* 45, no. 2 (2015): 161–86.

27 A. Chaix et al., "Time-Restricted Feeding Is a Preventative and Therapeutic Intervention against Diverse Nutritional Challenges," *Cell Metabolism* 20, no. 6 (2014): 991–1005.

28 T. Moro et al., "Effects of Eight Weeks of Time-Restricted Feeding (16/8) on Basal Metabolism, Maximal Strength, Body Composition, Inflammation, and Cardiovascular Risk Factors in Resistance-Trained Males," *Journal of Translational Medicine* 14 (2016): article no. 290.

29 P. Puchalska and P. A. Crawford, "Multi-Dimensional Roles of Ketone Bodies in Fuel Metabolism, Signaling, and Therapeutics," *Cell Metabolism* 25, no. 2 (2017): 262–84.

30 King, Burley, and Blundell, "Exercise-Induced Suppression."

CHAPTER 8: ADAPTING TO THE ULTIMATE DISRUPTERS: LIGHTS AND SCREENS

1 R. M. Lunn et al., "Health Consequences of Electric Lighting Practices in the Modern World: A Report on the National Toxicology Program's Workshop on Shift

Work at Night, Artificial Light at Night, and Circadian Disruption," *Science of Total Environment* 607–8 (2017): 1073–84.

2 C. A. Czeisler et al., "Bright Light Induction of Strong (Type 0) Resetting of the Human Circadian Pacemaker," *Science* 244, no. 4910 (1989): 1328–33.

3 J. Xu et al., "Altered Activity-Rest Patterns in Mice with a Human Autosomal-Dominant Nocturnal Frontal Lobe Epilepsy Mutation in the β2 Nicotinic Receptor," *Molecular Psychiatry* 16, no. 10 (2011): 1048–61.

4 L. A. Kirkby and M. B. Feller, "Intrinsically Photosensitive Ganglion Cells Contribute to Plasticity in Retinal Wave Circuits," *Proceedings of the National Academy of Sciences of the United States of America* 110, no. 29 (2013): 12090–95.

5 J. M. Renna, S. Weng, and D. M. Berson, "Light Acts through Melanopsin to Alter Retinal Waves and Segregation of Retinogeniculate Afferents," *Nature Neuroscience* 14, no. 7 (2011): 827–29.

6 J. Parent, W. Sanders, and R. Forehand, "Youth Screen Time and Behavioral Health Problems: The Role of Sleep Duration and Disturbances," *Journal of Developmental and Behavioral Pediatrics* 37, no. 4 (2016): 277–84.

7 *The Nielsen Total Audience Report: Q2 2017,* http://www.nielscn.com/us/en/insights/reports/2017/the-nielsen-total-audience-q2-2017.html.

8 I. Provencio et al., "Melanopsin: An Opsin in Melanophores, Brain, and Eye," *Proceedings of the National Academy of Sciences of the United States of America* 95, no. 1 (1998): 340–45.

9 P. A. Good, R. H. Taylor, and M. J. Mortimer, "The use of tinted glasses in childhood migraine." *Headache* 31 (1991): 533–536.

10 S. Vásquez-Ruiz et al., "A Light/Dark Cycle in the NICU Accelerates Body Weight Gain and Shortens Time to Discharge in Preterm Infants," *Early Human Development* 90, no. 9 (2014): 535–40.

11 P. A. Regidor et al., "Identification and Prediction of the Fertile Window with a New Web-Based Medical Device Using a Vaginal Biosensor for Measuring the Circadian and Circamensual Core Body Temperature," *Gynecological Endocrinology* 34, no. 3 (2018): 256–60.

12 X. Li et al., "Digital Health: Tracking Physiomes and Activity Using Wearable Biosensors Reveals Useful Health-Related Information," *PLoS Biology* 15, no. 1 (2017): e2001402.

13 C. Skarke et al., "A Pilot Characterization of the Human Chronobiome," *Scientific Reports* 7 (2017): article no. 17141.

14 D. Zeevi et al., "Personalized Nutrition by Prediction of Glycemic Responses," *Cell* 163, no. 5 (2015): 1079–94.

CHAPTER 9: THE CLOCK, THE MICROBIOME, AND DIGESTIVE CONCERNS

1 J. G. Moore, "Circadian Dynamics of Gastric Acid Secretion and Pharmacodynamics of H2 Receptor Blockade," *Annals of the New York Academy of Sciences* 618 (1991): 150–58.

2 K. Spiegel et al., "Brief Communication: Sleep Curtailment in Healthy Young Men Is Associated with Decreased Leptin Levels, Elevated Ghrelin Levels, and Increased Hunger and Appetite," *Annals of Internal Medicine* 141, no. 11 (2004): 846–50.

3 S. Taheri et al., "Short Sleep Duration Is Associated with Reduced Leptin, Elevated Ghrelin, and Increased Body Mass Index," *PLoS Medicine* 1, no. 3 (2004): e62.

4 J. Bradwejn, D. Koszycki, and G. Meterissian, Cholecystokinin-tetrapeptide Induces Panic Attacks in Patients with Panic Disorder. *Can J Psychiatry* 35 (1990): 83–85.

5 L. M. Ubaldo-Reyes, R. M. Buijs, C. Escobar, and M. Angeles-Castellanos, "Scheduled Meal Accelerates Entrainment to a 6-H Phase Advance by Shifting Central and Peripheral Oscillations in Rats," *European Journal of Neuroscience* 46, no. 3 (2017): 1875–86.

6 C. A. Thaiss et al., "Transkingdom Control of Microbiota Diurnal Oscillations Promotes Metabolic Homeostasis," *Cell* 159, no. 3 (2014): 514–29.

7 P. J. Turnbaugh et al., "Diet-Induced Obesity is Linked to Marked but Reversible Alterations in the Mouse Distal Gut Microbiome," *Cell Host & Microbe* 3, no. 4 (2008): 213–23.

8 Thaiss, "Transkingdom Control of Microbiota Diurnal Oscillations."

9 A. Zarrinpar et al., "Diet and Feeding Pattern Affect the Diurnal Dynamics of the Gut Microbiome," *Cell Metabolism* 20, no. 6 (2014): 1006–17.

10 J. A. Foster and K. A. McVey Neufeld, "Gut-Brain Axis: How the Microbiome Influences Anxiety and Depression," *Trends in Neurosciences* 36, no. 5 (2013): 305–12.

11 D. Hranilovic et al., "Hyperserotonemia in Adults with Autistic Disorder," *Journal of Autism and Developmental Disorders* 37, no. 10 (2007): 1934–40.

12 D. F. MacFabe et al., "Effects of the Enteric Bacterial Metabolic Product Propionic Acid on Object-Directed Behavior, Social Behavior, Cognition, and Neuroinflammation in Adolescent Rats: Relevance to Autism Spectrum Disorder," *Behavioural Brain Research* 217, no. 1 (2011): 47–54.

13 B. Chassaing et al., "Dietary Emulsifiers Impact the Mouse Gut Microbiota Promoting Colitis and Metabolic Syndrome," *Nature* 519, no. 7541 (2015): 92–96.

14 B. Chassaing et al., "Dietary Emulsifiers Directly Alter Human Microbiota Composition and Gene Expression Ex Vivo Potentiating Intestinal Inflammation," *Gut* 66, no. 8 (2017): 1414–27.

15 M. S. Desai et al., "A Dietary Fiber–Deprived Gut Microbiota Degrades the Colonic Mucus Barrier and Enhances Pathogen Susceptibility," *Cell* 167, no. 5 (2016): 1339–53.

16 K. Segawa et al., "Peptic Ulcer Is Prevalent among Shift Workers," *Digestive Diseases and Sciences* 32, no. 5 (1987): 449–53.

17 R. Shaker et al., "Nighttime Heartburn Is an Under-Appreciated Clinical Problem That Impacts Sleep and Daytime Function: The Results of a Gallup Survey Conducted on Behalf of the American Gastroenterological Association," *American Journal of Gastroenterology* 98, no. 7 (2003): 1487–93.

18 J. Leonard, J. K. Marshall, and P. Moayyedi, "Systematic Review of the Risk of Enteric Infection in Patients Taking Acid Suppression," *American Journal of Gastroenterology* 102, no. 9 (2007): 2047–56.

19 R. J. Hassing et al., "Proton Pump Inhibitors and Gastroenteritis," *European Journal of Epidemiology* 31, no. 10 (2016): 1057–63.

20 T. Antoniou et al., "Proton Pump Inhibitors and the Risk of Acute Kidney Injury in Older Patients: A Population-Based Cohort Study," *CMAJ Open* 3, no. 2 (2015): E166–71.

21 M. L. Blank et al., "A Nationwide Nested Case-Control Study Indicates an Increased Risk of Acute Interstitial Nephritis with Proton Pump Inhibitor Use," *Kidney International* 86, no. 4 (2014): 837–44.

22 P. Malfertheiner, A. Kandulski, and M. Venerito, "Proton-Pump Inhibitors: Understanding the Complications and Risks," *Nature Reviews: Gastroenterology & Hepatology* 14, no. 12 (2017): 697–710.

23 T. Ito and R. T. Jensen, "Association of Long-Term Proton Pump Inhibitor Therapy with Bone Fractures and Effects on Absorption of Calcium, Vitamin B12, Iron, and Magnesium," *Current Gastroenterology Reports* 12, no. 6 (2010): 448–57.

CHAPTER 10: THE CIRCADIAN CODE FOR ADDRESSING METABOLIC SYNDROME: OBESITY, DIABETES, AND HEART DISEASE

1 National Institute of Diabetes and Digestive and Kidney Diseases, "Health Risks of Being Overweight," https://www.niddk.nih.gov/health-information/weight-management/health-risks-overweight.

2 Y. Ma et al., "Association Between Eating Patterns and Obesity in a Free-Living US Adult Population," *American Journal of Epidemiology* 158, no. 1 (2003): 85–92.

3 A. K. Kant and B. I. Graubard, "40-Year Trends in Meal and Snack Eating Behaviors of American Adults," *Journal of the Academy of Nutrition and Dietetics* 115, no. 1 (2015): 50–63.

4 S. Gill and S. Panda, "A Smartphone App Reveals Erratic Diurnal Eating Patterns in Humans That Can Be Modulated for Health Benefits," *Cell Metabolism* 22, no. 5 (2015): 789–98.

5 N. J. Gupta, V. Kumar, and S. Panda, "A Camera-Phone Based Study Reveals Erratic Eating Pattern and Disrupted Daily Eating-Fasting Cycle among Adults in India," *PLoS ONE* 12, no. 3 (2017): e0172852.

6 A. J. Stunkard, W. J. Grace, and H. G. Wolff, "The Night-Eating Syndrome: A Pattern of Food Intake among Certain Obese Patients," *American Journal of Medicine* 19, no. 1 (1955): 78–86.

7 E. Takeda et al., "Stress Control and Human Nutrition," *Journal of Medical Investigation* 51, no. 3–4 (2004): 139–45.

8 Z. Liu et al., "PER1 Phosphorylation Specifies Feeding Rhythm in Mice," *Cell Reports* 7, no. 5 (2014): 1509–20.

9 T. Tuomi et al., "Increased Melatonin Signaling Is a Risk Factor for Type 2 Diabetes," *Cell Metabolism* 23, no. 6 (2016): 1067–77.

10 M. Watanabe et al., "Bile Acids Induce Energy Expenditure by Promoting Intracellular Thyroid Hormone Activation," *Nature* 439, no. 7075 (2006): 484–89.

11 A. Chaix et al., "Time-Restricted Feeding Is a Preventative and Therapeutic Intervention against Diverse Nutritional Challenges," *Cell Metabolism* 20, no. 6 (2014): 991–1005.

12 P. N. Hopkins, "Molecular Biology of Atherosclerosis," *Physiological Reviews* 93, no. 3 (2013): 1317–1542.

13 D. Montaigne et al., "Daytime Variation of Perioperative Myocardial Injury in Cardiac Surgery and Its Prevention by Rev-Erbα Antagonism: A Single-Centre Propensity-Matched Cohort Study and a Randomised Study," *Lancet* 391, no. 10115 (2017): 59–69.

CHAPTER II: ENHANCING THE IMMUNE SYSTEM AND TREATING CANCER

1 C. N. Bernstein et al., "Cancer Risk in Patients with Inflammatory Bowel Disease: A Population-Based Study," *Cancer* 91, no. 4 (2001): 854–62.

2 N. B. Milev and A. B. Reddy, "Circadian Redox Oscillations and Metabolism," *Trends in Endocrinology and Metabolism* 26, no. 8 (2015): 430–37.

3 N. Martinez-Lopez et al., "System-Wide Benefits of Internal Fasting by Autophagy," *Cell Metabolism* 26, no. 6 (2017): 856–71.

4 D. Cai et al., "Local and Systemic Insulin Resistance Resulting from Hepatic Activation of IKK-beta and NF-kappaB," *Nature Medicine* 11, no. 2 (2005): 183–90.

5 R. Narasimamurthy et al., "Circadian Clock Protein Cryptochrome Regulates the Expression of Proinflammatory Cytokines," *Proceedings of the National Academy of Sciences of the United States of America* 109, no. 31 (2012): 12662–67.

6 T. D. Girard et al., "Delirium as a Predictor of Long-Term Cognitive Impairment in Survivors of Critical Illness," *Critical Care Medicine* 38, no. 7 (2010): 1513–20.

7 S. Arumugam et al., "Delirium in the Intensive Care Unit," *Journal of Emergencies, Trauma, and Shock* 10, no. 1 (2017): 37–46.

8 B. van Rompaey et al., "The Effect of Earplugs during the Night on the Onset of Delirium and Sleep Perception: A Randomized Controlled Trial in Intensive Care Patients," *Critical Care* 16, no. 3 (2012): article no. R73.

9 A. Reinberg and F. Levi, "Clinical Chronopharmacology with Special Reference to NSAIDs," *Scandinavian Journal of Rheumatology: Supplement* 65 (1987): 118–22.

10 I. C. Chikanza, "Defective Hypothalamic Response to Immune and Inflammatory Stimuli in Patients with Rheumatoid Arthritis," *Arthritis Rheumatism* 35, no. 11 (1992): 1281–88.

11 F. Buttgereit et al., "Efficacy of Modified-Release versus Standard Prednisone to Reduce Duration of Morning Stiffness of the Joints in Rheumatoid Arthritis (CAPRA-1): A Double-Blind, Randomised Controlled Trial," *Lancet* 371, no. 9608 (2008): 205–14.

12 A. Ballesta et al., "Systems Chronotherapeutics," *Pharmacological Reviews* 69, no. 2 (2017): 161–99.

13 K. Spiegel, J. F. Sheridan, and E. van Cauter, "Effect of Sleep Deprivation on Response to Immunization," *JAMA: The Journal of the American Medical Association* 288, no. 12 (2002): 1471–72.

14 J. E. Long et al., "Morning Vaccination Enhances Antibody Response over After-noon Vaccination: A Cluster-Randomised Trial," *Vaccine* 34, no. 24 (2016): 2679–85.

15 O. Castanon-Cervantes, "Dysregulation of Inflammatory Responses by Chronic Circadian Disruption," *Journal of Immunology* 185, no. 10 (2010): 5796–805.

16 Y. M. Cissé et al., "Time-Restricted Feeding Alters the Innate Immune Response to Bacterial Endotoxin," *Journal of Immunology* 200, no. 2 (2018): 681–87.

17 J. Samulin Erdem et al., "Mechanisms of Breast Cancer Risk in Shift Workers: Association of Telomere Shortening with the Duration and Intensity of Night Work," *Cancer Medicine* 6, no. 8 (2017): 1988–97.

18 C. R. Marinac et al., "Prolonged Nightly Fasting and Breast Cancer Risk: Findings from NHANES (2009–2010)," *Cancer Epidemiology, Biomarkers & Prevention* 24, no. 5 (2015): 783–89.

19 E. Filipski et al., "Effects of Light and Food Schedules on Liver and Tumor Molecular Clocks in Mice," *Journal of the National Cancer Institute* 97, no. 7 (2005): 507–17.

20 M. W. Wu et al., "Effects of Meal Timing on Tumor Progression in Mice," *Life Sciences* 75, no. 10 (2004): 1181–93.

21 W. J. Hrushesky, "Circadian Timing of Cancer Chemotherapy," *Science* 228, no. 4695 (1985): 73–75.

22 R. Dallmann, A. Okyar, and F. Levi, "Dosing-Time Makes the Poison: Circadian Regulation and Pharmacotherapy," *Trends in Molecular Medicine* 22, no. 5 (2016): 430–35.

23 F. Levi et al., "Oxaliplatin Activity Against Metastatic Colorectal Cancer. A Phase II Study of 5-Day Continuous Venous Infusion at Circadian Rhythm Modulated Rate," *European Journal of Cancer* 29A, no. 9 (1993): 1280–84.

24 T. Matsuo et al., "Control Mechanism of the Circadian Clock for Timing of Cell Division In Vivo," *Science* 302, no. 5643 (2003): 255–59.

25 M. V. Plikus et al., "Local Circadian Clock Gates Cell Cycle Progression of Transient Amplifying Cells during Regenerative Hair Cycling," *Proceedings of the National Academy of Sciences of the United States of America* 110, no. 23 (2013): E2106–15.

26 S. Kiessling et al., "Enhancing Circadian Clock Function in Cancer Cells Inhibits Tumor Growth," *BMC Biology* 15 (2017): article no. 13.

27 G. Sulli et al., "Pharmacological Activation of REV-ERBs Is Lethal in Cancer and Oncogene-Induced Senescence," *Nature* 553, no. 7688 (2018): 351–55.

28 J. Marescaux et al., "Transatlantic Robot-Assisted Telesurgery," *Nature* 413, no. 6854 (2001): 379–80.

29 J. Marescaux et al., "Transcontinental Robot-Assisted Remote Telesurgery: Feasibility and Potential Applications," *Annals of Surgery* 235, no. 4 (2002): 487–92.

30 C. R. Marinac et al., "Prolonged Nightly Fasting and Breast Cancer Prognosis," *JAMA Oncology* 2, no. 8 (2016): 1049–55.

CHAPTER 12: THE CIRCADIAN CODE FOR OPTIMIZING BRAIN HEALTH

1 P. S. Eriksson et al., "Neurogenesis in the Adult Human Hippocampus," *Nature Medicine* 4, no. 11 (1998): 1313–17.

2 R. Noseda et al., "A Neural Mechanism for Exacerbation of Headache by Light," *Nature Neuroscience* 13, no. 2 (2010): 239–45.

3 J. Kim et al., "Implications of Circadian Rhythm in Dopamine and Mood Regulation," *Molecules and Cells* 40, no. 7 (2017): 450–56.

4 G. E. Davis and W. E. Lowell, "Evidence That Latitude Is Directly Related to Variation in Suicide Rates," *Canadian Journal of Psychiatry* 47, no. 6 (2002): 572–74.

5 T. Terao et al., "Effect of Latitude on Suicide Rates in Japan," *Lancet* 360, no. 9348 (2002): 1892.

6 C. L. Drake et al., "Shift Work Sleep Disorder: Prevalence and Consequences beyond That of Symptomatic Day Workers," *Sleep* 27, no. 8 (2004): 1453–62.

7 A. Azzi et al., "Network Dynamics Mediate Circadian Clock Plasticity," *Neuron* 93, no. 2 (2017): 441–50.

8 A. Azzi et al., "Circadian Behavior Is Light-Reprogrammed by Plastic DNA Methylation," *Nature Neuroscience* 17, no. 3 (2014): 377–82.

9 C. J. Madrid-Navarro et al., "Disruption of Circadian Rhythms and Delirium, Sleep Impairment and Sepsis in Critically Ill Patients: Potential Therapeutic Implications for Increased Light-Dark Contrast and Melatonin Therapy in an ICU Environment," *Current Pharmaceutical Design* 21, no. 24 (2015): 3453–68.

10 S. Vásquez-Ruiz et al., "A Light/Dark Cycle in the NICU Accelerates Body Weight Gain and Shortens Time to Discharge in Preterm Infants," *Early Human Development* 90, no. 9 (2014): 535–40.

11 K. Wulff et al., "Sleep and Circadian Rhythm Disruption in Psychiatric and Neurodegenerative Disease," *Nature Reviews: Neuroscience* 11, no. 8 (2010): 589–99.

12 L. Xie et al., "Sleep Drives Metabolite Clearance from the Adult Brain," *Science* 342, no. 6156 (2013): 373–77.

13 J. Mattis and A. Sehgal, "Circadian Rhythms, Sleep, and Disorders of Aging," *Trends in Endocrinology and Metabolism* 27, no. 4 (2016): 192–203.

14 J. E. Kang et al., "Amyloid-β Dynamics Are Regulated by Orexin and the Sleep-Wake Cycle," *Science* 326, no. 5955 (2009): 1005–7.

15 A. Di Meco, Y. B. Joshi, and D. Pratico, "Sleep Deprivation Impairs Memory, Tau Metabolism, and Synaptic Integrity of a Mouse Model of Alzheimer's Disease with Plaques and Tangles," *Neurobiology of Aging* 35, no. 8 (2014): 1813–20.

16 J. Vienne et al., "Age-Related Reduction of Recovery Sleep and Arousal Threshold in *Drosophila*," *Sleep* 39, no. 8 (2016): 1613–24.

17 A. Chaix and S. Panda, "Ketone Bodies Signal Opportunistic Food-Seeking Activity," *Trends in Endocrinology & Metabolism* 27, no. 6 (2016): 350–52.

18 R. Chavan et al., "Liver-Derived Ketone Bodies Are Necessary for Food Anticipation," *Nature Communications* 7 (2016): article no. 10580.

19 M. P. Mattson, "Lifelong Brain Health Is a Lifelong Challenge: From Evolutionary Principles to Empirical Evidence," *Ageing Research Reviews* 20 (2015): 37–45.

20 H. B. Wang et al., "Time-Restricted Feeding Improves Circadian Dysfunction as Well as Motor Symptoms in the Q175 Mouse Model of Huntington's Disease," *eNeuro* 5, no. 1 (2018): doi: 10.1523/ENEURO.0431-17.2017.

21 M. C. Yoon et al., "Treadmill Exercise Suppresses Nigrostriatal Dopaminergic Neuronal Loss in 6-Hydroxydopamine-Induced Parkinson's Rats," *Neuroscience Letters* 423, no. 1 (2007): 12–17.

22 C. W. Cotman, N. C. Berchtold, and L. A. Christie, "Exercise Builds Brain Health: Key Roles of Growth Factor Cascades and Inflammation," *Trends in Neurosciences* 30, no. 9 (2007): 464–72.

23 A. J. Bruce-Keller et al., "Food Restriction Reduces Brain Damage and Improves Behavioral Outcome Following Excitotoxic and Metabolic Insults," *Annals of Neurology* 45, no. 1 (1999): 8–15.

24 M. L. Inder, M. T. Crowe, and R. Porter, "Effect of Transmeridian Travel and Jetlag on Mood Disorders: Evidence and Implications," *Australian and New Zealand Journal of Psychiatry* 50, no. 3 (2016): 220–27.

25 L. Yin et al., "Nuclear Receptor Rev-erbα Is a Critical Lithium-Sensitive Component of the Circadian Clock," *Science* 311, no. 5763 (2006): 1002–5.

26 Emily Manoogian, "A Prized Life: A Glimpse into the Life of Nobel Laureate, Dr. Roger Guillemin," *myCircadianClock* (blog), May 6, 2016, http://blog.mycircadian clock.org/a-prized-life-a-glimpse-into-the-life-of-nobel-laureate-dr-roger-guille min/.

INDEX